Cambridge IGCSE™
Mathematics
CORE PRACTICE BOOK

Karen Morrison

CAMBRIDGE
UNIVERSITY PRESS

Shaftesbury Road, Cambridge CB2 8EA, United Kingdom

One Liberty Plaza, 20th Floor, New York, NY 10006, USA

477 Williamstown Road, Port Melbourne, VIC 3207, Australia

314–321, 3rd Floor, Plot 3, Splendor Forum, Jasola District Centre, New Delhi – 110025, India

103 Penang Road, #05–06/07, Visioncrest Commercial, Singapore 238467

Cambridge University Press is part of the University of Cambridge.

It furthers the University's mission by disseminating knowledge in the pursuit of education, learning and research at the highest international levels of excellence.

www.cambridge.org
Information on this title: www.cambridge.org/9781009297950

© Cambridge University Press & Assessment 2023

This publication is in copyright. Subject to statutory exception and to the provisions of relevant collective licensing agreements, no reproduction of any part may take place without the written permission of Cambridge University Press.

First published 2012
Second edition 2018
Third edition 2023

20 19 18 17 16 15 14 13 12 11 10 9 8 7 6 5 4 3 2 1

Printed in Italy by L.E.G.O. S.p.A.

A catalogue record for this publication is available from the British Library

ISBN 978-1-009-29795-0 Paperback with Digital Access (2 Years)

Additional resources for this publication at www.cambridge.org/go

Cambridge University Press has no responsibility for the persistence or accuracy of URLs for external or third-party internet websites referred to in this publication, and does not guarantee that any content on such websites is, or will remain, accurate or appropriate. Information regarding prices, travel timetables, and other factual information given in this work is correct at the time of first printing but Cambridge University Press does not guarantee the accuracy of such information thereafter.

....

NOTICE TO TEACHERS IN THE UK
It is illegal to reproduce any part of this work in material form (including photocopying and electronic storage) except under the following circumstances:
(i) where you are abiding by a licence granted to your school or institution by the Copyright Licensing Agency;
(ii) where no such licence exists, or where you wish to exceed the terms of a licence, and you have gained the written permission of Cambridge University Press;
(iii) where you are allowed to reproduce without permission under the provisions of Chapter 3 of the Copyright, Designs and Patents Act 1988, which covers, for example, the reproduction of short passages within certain types of educational anthology and reproduction for the purposes of setting examination questions.

CAMBRIDGE DEDICATED TEACHER AWARDS 2022

Teachers play an important part in shaping futures. Our Dedicated Teacher Awards recognise the hard work that teachers put in every day.

Thank you to everyone who nominated this year; we have been inspired and moved by all of your stories. Well done to all of our nominees for your dedication to learning and for inspiring the next generation of thinkers, leaders and innovators.

Congratulations to our incredible winners!

WINNER

Regional Winner — Australia, New Zealand & South-East Asia
Mohd Al Khalifa Bin Mohd Affnan
Keningau Vocational College, Malaysia

Regional Winner — Europe
Dr. Mary Shiny Ponparambil Paul
Little Flower English School, Italy

Regional Winner — North & South America
Noemi Falcon
Zora Neale Hurston Elementary School, United States

Regional Winner — Central & Southern Africa
Temitope Adewuyi
Fountain Heights Secondary School, Nigeria

Regional Winner — Middle East & North Africa
Uroosa Imran
Beaconhouse School System KG-1 branch, Pakistan

Regional Winner — East & South Asia
Jeenath Akther
Chittagong Grammar School, Bangladesh

For more information about our dedicated teachers and their stories, go to
dedicatedteacher.cambridge.org

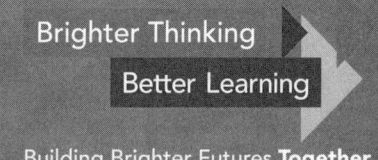

Building Brighter Futures Together

Endorsement statement

Endorsement indicates that a resource has passed Cambridge International's rigorous quality-assurance process and is suitable to support the delivery of a Cambridge International syllabus. However, endorsed resources are not the only suitable materials available to support teaching and learning, and are not essential to be used to achieve the qualification. Resource lists found on the Cambridge International website will include this resource and other endorsed resources.

Any example answers to questions taken from past question papers, practice questions, accompanying marks and mark schemes included in this resource have been written by the authors and are for guidance only. They do not replicate examination papers. In examinations the way marks are awarded may be different. Any references to assessment and/or assessment preparation are the publisher's interpretation of the syllabus requirements. Examiners will not use endorsed resources as a source of material for any assessment set by Cambridge International.

While the publishers have made every attempt to ensure that advice on the qualification and its assessment is accurate, the official syllabus, specimen assessment materials and any associated assessment guidance materials produced by the awarding body are the only authoritative source of information and should always be referred to for definitive guidance. Cambridge International recommends that teachers consider using a range of teaching and learning resources based on their own professional judgement of their students' needs.

Cambridge International has not paid for the production of this resource, nor does Cambridge International receive any royalties from its sale. For more information about the endorsement process, please visit www.cambridgeinternational.org/endorsed-resources

Cambridge International copyright material in this publication is reproduced under licence and remains the intellectual property of Cambridge Assessment International Education.

Third party websites and resources referred to in this publication have not been endorsed by Cambridge Assessment International Education.

Contents

Introduction		viii
How to use this book		ix
How to use this series		x

1 Review of number concepts — 1
- 1.1 Different types of numbers — 1
- 1.2 Multiples and factors — 2
- 1.3 Prime numbers — 3
- 1.4 Working with directed numbers — 4
- 1.5 Powers, roots and laws of indices — 5
- 1.6 Order of operations — 7
- 1.7 Rounding and estimating — 8

2 Making sense of algebra — 10
- 2.1 Using letters to represent unknown values — 10
- 2.2 Substitution — 11
- 2.3 Simplifying expressions — 12
- 2.4 Working with brackets — 13
- 2.5 Indices — 14

3 Lines, angles and shapes — 18
- 3.1 Lines and angles — 18
- 3.2 Triangles — 21
- 3.3 Quadrilaterals — 23
- 3.4 Polygons — 25
- 3.5 Circles — 26
- 3.6 Construction — 27

4 Collecting, organising and displaying data — 30
- 4.1 Collecting and classifying data — 30
- 4.2 Organising data — 31
- 4.3 Using charts to display data — 34

5 Fractions, percentages and standard form — 40
- 5.1 Revisiting fractions — 40
- 5.2 Operations on fractions — 40
- 5.3 Percentages — 42
- 5.4 Standard form — 45

6 Equations, factors and formulae — 48
- 6.1 Solving equations — 48
- 6.2 Factorising algebraic expressions — 49
- 6.3 Rearranging formula — 50

7 Perimeter, area and volume — 53
- 7.1 Perimeter and area in two dimensions — 53
- 7.2 Three-dimensional objects — 58
- 7.3 Surface areas and volumes of solids — 59

8 Introduction to probability — 66
- 8.1 Understanding basic probability — 66
- 8.2 Sample space diagrams — 69
- 8.3 Combining independent and mutually exclusive events — 70

9 Sequences and sets — 73
- 9.1 Sequences — 73
- 9.2 Rational and irrational numbers — 75
- 9.3 Sets — 75

10 Straight lines and quadratic expressions — 79
- 10.1 Straight line graphs — 79
- 10.2 Quadratic expressions — 82

11 Pythagoras' theorem and similar shapes — 86
- 11.1 Pythagoras' theorem — 86
- 11.2 Understanding similar triangles — 88
- 11.3 Understanding similar shapes — 90
- 11.4 Understanding congruence — 91

12 Averages and measures of spread — 94
- 12.1 Different types of average — 94
- 12.2 Making comparisons using averages and ranges — 97
- 12.3 Calculating averages and ranges for frequency data — 98

13 Understanding measurement — 102
- 13.1 Understanding units — 102
- 13.2 Time — 104
- 13.3 Limits of accuracy – upper and lower bounds — 105
- 13.4 Conversion graphs — 106
- 13.5 Exchanging currencies — 108

14 Further solving of equations and inequalities — 111
- 14.1 Simultaneous linear equations — 111
- 14.2 Linear inequalities — 114

15 Scale drawings, bearings and trigonometry — 117
- 15.1 Scale drawings — 117
- 15.2 Bearings — 118
- 15.3 Understanding the tangent, cosine and sine ratios — 120
- 15.4 Solving problems using trigonometry — 122

16 Scatter diagrams and correlation — 126
- 16.1 Introduction to bivariate data — 126

17 Managing money — 130
- 17.1 Earning money — 130
- 17.2 Borrowing and investing money — 131
- 17.3 Buying and selling — 132

18 Curved graphs — 135
- 18.1 Review of quadratic graphs (the parabola) — 135
- 18.2 Drawing reciprocal graphs (the hyperbola) — 137
- 18.3 Using graphs to solve quadratic equations — 138

19 Symmetry — 141
- 19.1 Symmetry in two dimensions — 141
- 19.2 Angle relationships in circles — 142

20 Ratio, rate and proportion — 145
- 20.1 Working with ratio — 145
- 20.2 Ratio and scale — 146
- 20.3 Rates — 147
- 20.4 Kinematic graphs — 148
- 20.5 Proportion — 150

21 More equations and formulae — 154
- 21.1 Setting up equations to solve problems — 154

22 Transformations — 157
- 22.1 Simple plane transformations — 157

23 Probability using tree diagrams and Venn diagrams — 164

- 23.1 Using tree diagrams to show outcomes — 164
- 23.2 Calculating probability from tree diagrams — 164
- 23.3 Calculating probability from Venn diagrams — 166

Acknowledgements — 168

Introduction

Have you heard the saying 'practice makes perfect' or read articles about how musicians, athletes and other people become experts in their fields by putting in 10 000 hours of practice? Practising basic skills and techniques can help you improve in many areas of life, including mathematics.

This book offers plenty of practice opportunities for each topic in the syllabus. Once you have read through the summary of key concepts in each topic, you can work through the basic 'drill and practice' questions to perfect your skills and techniques. The answers for all questions are available to download from Cambridge GO so that you can check your own work and make corrections or do revision as you need to.

Once you are confident in the basics, you can work though the 'Review exercise' at the end of each chapter. The review exercises bring together the sub-topics in each chapter and give you the chance to apply the skills you have practised and to help you choose the techniques and skills that are most suited to solve more structured and integrated problems. Again, you can find the answers on Cambridge GO to check your own work.

In line with modern thinking about how people learn best, we've built in opportunities for different types of self-assessment. The feedback from these assessments will help you decide what revision or practice you can do to improve your understanding and/or performance.

There are also many opportunities for you to reflect on your own learning style and to think about what you can do really well and to consider what you can learn from your mistakes. The reflection questions aim to encourage a growth mindset and develop positive attitudes towards learning maths.

We hope that you will enjoy working through the course and that you will find the material interesting, engaging and worthwhile.

Karen Morrison

> How to use this book

Throughout this book, you will notice lots of different features that will help your learning. These are explained below.

Exercises

These help you to practise skills that are important for studying Cambridge IGCSE™ Mathematics. There are two types of exercise:

- Exercises which let you practise the mathematical skills you have learned.

- Review exercises which bring together all the mathematical concepts in a chapter, pushing your skills further.

KEY LEARNING STATEMENT

This will remind you of what you should already know from your previous study in order to complete the exercises in this section.

REFLECTION

At the end of some exercises you will find opportunities to think about the approach that you take to your work, and how you might improve this in the future.

SELF ASSESSMENT

At the end of some exercises, you will find opportunities to help you assess your own work, and consider how you can improve the way you learn.

KEY CONCEPTS

These summarise the important concepts that are covered in each section.

TIP

The information in these boxes will help you complete the exercises, and give you support in areas that you might find difficult.

 This icon shows you where you should complete an exercise without using your calculator.

CAMBRIDGE IGCSE™ MATHEMATICS: CORE PRACTICE BOOK

> How to use this series

This suite of resources supports learners and teachers following the Cambridge IGCSE™ and IGCSE (9–1) Mathematics syllabuses (0580/0980). Up-to-date metacognition techniques have been incorporated throughout the resources to meet the changes in the syllabus and develop a complete understanding of mathematics for learners. All of the components in the series are designed to work together.

The coursebook contains six units that together offer complete coverage of the syllabus. We have worked with NRICH to provide a variety of new project activities, designed to engage learners and strengthen their problem-solving skills. A new Mathematical Connections feature creates a holistic view of mathematics to help learners identify links between themes and topics. Each chapter contains opportunities for formative assessment, differentiation and peer and self-assessment offering learners the support needed to make progress. Cambridge Online Mathematics is available through the digital/print bundle option or on its own without the print coursebook. Learners can review content digitally, explore worked examples and test their knowledge with practice questions and answers. Teachers benefit from the ability to set tests and tasks with the added auto-marking functionality and a reporting dashboard to help track learner progress quickly and easily.

The digital teacher's resource provides extensive guidance on how to teach the course, including suggestions for differentiation, formative assessment and language support, teaching ideas and PowerPoints. The Teaching Skills Focus shows teachers how to incorporate a variety of key pedagogical techniques into teaching, including differentiation, assessment for learning, and metacognition. Answers for all components are accessible to teachers for free on the Cambridge GO platform.

There are two practice books available, one for the core content of the syllabus and the other for learners studying extended content. These resources, which can be used in class or assigned as homework, provide a wide variety of extra maths activities and questions to help learners consolidate their learning and prepare for assessment. 'Tips' are also regularly featured to give learners extra advice and guidance on the different areas of maths they encounter. Access to the digital versions of the practice books is included, and answers can be found online.

Chapter 1: Review of number concepts

1.1 Different types of numbers

> **KEY LEARNING STATEMENTS**
> - Real numbers are either rational or irrational.
> - You can write rational numbers as fractions in the form $\frac{a}{b}$ where a and b are integers and $b \neq 0$. (Integers are negative and positive whole numbers, and zero.)
> - Rational numbers include integers, fractions, recurring and terminating decimals and percentages.

> **KEY CONCEPTS**
> - Classifying and using different types of numbers.
> - Interpreting and using the symbols $=$, $\neq$, $<$, $>$, $\leq$ and $\geq$.

1 Copy and complete this table by writing a definition and giving an example of each type of number.

Mathematical name	Definition	Example
Natural numbers		
Integers		
Prime numbers		
Square numbers		
Fractions		

> **TIP**
> Knowing the correct mathematical terms is important for understanding questions and communicating mathematically.

2 Give an example to show what each of the following symbols means.

a $>$ b $\leq$ c $\therefore$ d $\sqrt{}$

e $\neq$ f $\geq$ g $<$

3 Look at this set of numbers.

> $3, -2, 0, 1, 9, 15, 4, 5, -7, 10, 32, -32, 21, 23, 25, 27, 29, \frac{1}{2}$

a Which of these numbers are **not** natural numbers?

b Which of these numbers are **not** integers?

c Which of these numbers are prime numbers?

d Which of these numbers are square numbers?

4 List:

   a four square numbers greater than 100

   b four rational numbers smaller than $\frac{1}{3}$

   c two prime numbers that are >80

   d the prime numbers <10.

5 Write each amount as a number.

   a Three hundred and sixty-five thousand, two hundred and eighty-nine.

   b One billion, seven hundred and three million, four hundred and seventy-three thousand, two hundred and twelve.

1.2 Multiples and factors

KEY LEARNING STATEMENTS

- When you multiply a number by another number you get a multiple of the original number.
- The lowest common multiple (LCM) of two numbers is the lowest number that is a multiple of both numbers.
- Any number that will divide into a number exactly is a factor of that number.
- The highest common factor (HCF) of two numbers is the highest number that is a factor of both numbers.

KEY CONCEPT

Finding the highest common factor and lowest common multiple of two numbers.

1 Find the LCM of the given numbers.

   a 9 and 18   b 12 and 18   c 15 and 18   d 24 and 12

   e 36 and 9   f 12 and 8   g 9 and 24   h 12 and 32

2 Find the HCF of the given numbers.

   a 12 and 18   b 18 and 36   c 27 and 90   d 12 and 15

   e 20 and 30   f 19 and 45   g 60 and 72   h 250 and 900

3 Amira has two rolls of cotton fabric. One roll has 72 metres on it and the other has 90 metres on it. She wants to cut the fabric to make as many equal length pieces as possible of the longest possible length. How long should each piece be?

TIP

To find the LCM of a set of numbers, you can list the multiples of each number until you find the first multiple that is in the lists for all of the numbers in the set.

TIP

You need to work out whether to use LCM or HCF to find the answers. Problems involving LCM usually include repeating events. Problems involving HCF usually involve splitting things into smaller pieces or arranging things in equal groups or rows.

4   In a shopping mall promotion every 30th shopper gets a $10 voucher and every 120th shopper gets a free meal. How many shoppers must enter the mall before one receives both a voucher and a free meal?

5   Amanda has 40 pieces of fruit and 100 sweets to share with the students in her class. She is able to give each student an equal number of pieces of fruit and an equal number of sweets. What is the largest possible number of students in her class?

6   Sam has sheets of green and yellow plastic that he wants to use to make a square chequerboard pattern on a coffee-table top. Each sheet measures 210 cm by 154 cm. The squares are to be the maximum size possible. What will be the length of the side of each square and how many will he be able to cut from each sheet?

> **REFLECTION**
>
> Read these problems carefully. How can they help you to recognise similar problems in future, even if you are not told to use HCF and LCM?

1.3 Prime numbers

> **KEY LEARNING STATEMENTS**
>
> - Prime numbers only have two factors: 1 and the number itself.
> - Prime factors are factors of a number that are also prime numbers.
> - You can write any number as a product of prime factors. Remember the number 1 itself is *not* a prime number, so you cannot use it to write a number as the product of its prime factors.
> - You can use the product of prime factors to find the HCF or LCM of two numbers.

> **KEY CONCEPT**
>
> Prime numbers and prime factors.

1   Identify the prime numbers in each set.

   a   1, 2, 3, 4, 5, 6, 7, 8, 9, 10

   b   50, 51, 52, 53, 54, 55, 56, 57, 58, 59, 60

   c   95, 96, 97, 98, 99, 100, 101, 102, 103, 104, 105

2   Express the following numbers as a product of their prime factors.

   a   36   b   65   c   64   d   84
   e   80   f   1000   g   1270   h   1963

3   Find the LCM and the HCF of the following numbers by using prime factors.

   a   27 and 14   b   85 and 15   c   96 and 27   d   53 and 16
   e   674 and 72   f   234 and 66 g   550 and 128 h   315 and 275

> **TIP**
>
> You can use a tree diagram or division to find the prime factors of a composite whole number.

1.4 Working with directed numbers

> **KEY LEARNING STATEMENTS**
>
> - Integers are directed whole numbers.
> - You write negative integers with a minus (−) sign. Positive integers may be written with a plus (+) sign, but usually they are not.
> - In real life, negative numbers are used to represent temperatures below zero; movements downwards or left; depths; distances below sea level; bank withdrawals and overdrawn amounts, and many more things.

> **KEY CONCEPTS**
>
> - Using directed numbers in practical situations.
> - Basic calculations with positive and negative numbers.

1 If the temperature is 4 °C in the evening and it drops 7 °C overnight, what will the temperature be in the morning?

2 Which is colder in each pair of temperatures?

  a 0 °C or −2 °C   **b** 9 °C or −9 °C   **c** −4 °C or −12 °C

3 An office block has three basement levels (−1, −2 and −3), a ground floor (0) and 15 floors above the ground floor (1 to 15). Where will the lift be in the following situations?

  a Starts on the ground floor and goes down one floor then up five?

  b Starts on level −3 and goes up ten floors?

  c Starts on floor 12 and goes down 13 floors?

  d Starts on floor 15 and goes down 17 floors?

  e Starts on level −2, goes up seven floors and then down eight?

4 Write the number that is 12 less than:

  a 9   **b** −14   **c** −2   **d** 12

5 Calculate:

  a $-400 \div 80$

  b $-54 + 120 + (-25)$

  c $-3 \times (14 - (-12))$

  d $\dfrac{-18}{6} \times 3$

  e $13 + (-7) + 25 + (-15)$

6 The table shows how much the value of a rupee changed in comparison to the euro over a period of five days. The rate was 80.72 rupees : 1 euro before any changes were recorded.

Day	1	2	3	4	5
Change	−0.25	+0.14	−0.27	−2.08	−3.04

  a What was the value of the rupee compared to the euro and the end of day 3?

  b What was the total change over the period of five days? Give your answer as a directed number.

1.5 Powers, roots and laws of indices

KEY LEARNING STATEMENTS

- Index notation is a way of writing repeated multiplication. For example, you can write 2 × 2 × 2 as 2^3. 2 is the base and 3 is the index that tells you how many times 2 is multiplied by itself.
- The $\sqrt[x]{n}$ of a number is the value that is multiplied by itself x times to reach that number.
- Any number to the power of 0 is equal to 1: $a^0 = 1$.
- Negative indices are used to write reciprocals. a^{-m} is the reciprocal of a^m because $a^{-m} \times a^m = 1$.
- To multiply numbers with the same base you add the indices. In general terms $a^m \times a^n = a^{m+n}$.
- To divide numbers with the same base you subtract the indices. In general terms $\dfrac{a^m}{a^n} = a^{m-n}$.
- To raise a power to another power you multiply the indices. In general terms $(a^m)^n = a^{mn}$.

KEY CONCEPTS

- Calculating with squares, square roots, cubes, cube roots and other powers and roots of numbers.
- The meaning of zero and negative indices.
- The laws of indices.

1 Calculate.

   a 3^2 **b** 18^2 **c** 21^2 **d** 25^2

   e 6^3 **f** 15^3 **g** 18^3 **h** 35^3

2 Find these roots.

   a $\sqrt{121}$ **b** $\sqrt[3]{512}$ **c** $\sqrt{441}$

   d $\sqrt[3]{1331}$ **e** $\sqrt[3]{46656}$ **f** $\sqrt{2601}$

   g $\sqrt{3136}$ **h** $\sqrt{729}$ **i** $\sqrt[4]{1296}$

3 Find all the square and cube numbers between 100 and 300.

4 Which of the following are square numbers and which are cube numbers?

> 1, 24, 49, 64, 256, 676, 625, 128

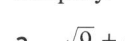

5 Simplify.

   a $\sqrt{9} + \sqrt{16}$ **b** $\sqrt{9+16}$ **c** $\sqrt{64} + \sqrt{36}$

   d $\sqrt{64+36}$ **e** $\sqrt{\dfrac{36}{4}}$ **f** $\left(\sqrt{25}\right)^2$

   g $\dfrac{\sqrt{9}}{\sqrt{16}}$ **h** $\sqrt{169-144}$ **i** $\sqrt[3]{27} - \sqrt[3]{1}$

   j $\sqrt{100 \div 4}$ **k** $\sqrt{1} + \sqrt{\dfrac{9}{16}}$ **l** $\sqrt{16} \times \sqrt[3]{27}$

TIP

If you don't have a calculator, you can use the product of prime factors to find the square root or cube root of a number.

6 Evaluate.

a 4^3 b 7^4

c 16^4 d 12^3

e 20^3 f 10^5

g $13^3 - 3^5$ h $3^3 + 2^7$

i $\sqrt[3]{64} + 4^5$ j $(2^4)^3$

7 Rewrite each of the following using only positive indices.

a 4^{-1} b 5^{-1}

c 8^{-1} d 5^{-2}

e 3^{-3} f 2^{-5}

g 3^{-4} h 8^{-6}

i 23^{-3} j 12^{-4}

8 Express each term using a negative index.

a $\dfrac{1}{2}$ b $\dfrac{1}{6}$

c $\dfrac{1}{3^2}$ d $\dfrac{1}{2^3}$

e $\dfrac{1}{3^3}$ f $\dfrac{1}{2^4}$

g $\dfrac{1}{11^2}$ h $\dfrac{1}{4^3}$

i $\dfrac{2}{10}$ j $\dfrac{3}{9}$

9 Simplify. Leave your answers in index form.

a $3^2 \times 3^6$ b $10^{-2} \times 10^4$

c $3^8 \times 3^{-5}$ d $5^0 \times 3^2$

e $2^{-3} \times 2^{-4}$ f $3 \times 3^2 \times 3^{-2}$

g $4^0 \times 4^{-2} \times 4$ h $10^2 \times 10^3 \times 10^{-2}$

i $(3^2)^0$ j $(4^3)^4$

k $(3^{-2})^{-3}$ l $(4^{-3} \times 4^2)^{-2}$

m $\dfrac{10^6}{10^{-3}}$ n $\dfrac{10^0}{10^4}$

o $\dfrac{2^{-4}}{2^{-5}}$ p $\dfrac{4^3}{4^{-3}}$

1.6 Order of operations

KEY LEARNING STATEMENTS

- When there is more than one operation to be done in a calculation you must work out the parts in brackets first. Next deal with powers and roots. Then do any division or multiplication (from left to right) before adding and subtracting (from left to right).

- Long fraction lines and square or cube root signs act like brackets, indicating parts of the calculation that have to be done first.

- Scientific calculators apply the rules for order of operations automatically. If there are brackets, fractions or roots in your calculation you need to enter these correctly on the calculator. When there is more than one term in the denominator, the calculator will divide by the first term only unless you enter brackets.

KEY CONCEPT

Calculating using the correct order of operations.

1 Calculate and give your answer correct to two decimal places.

- **a** $8 + 3 \times 6$
- **b** $(8 + 3) \times 6$
- **c** $8 \times 3 - 4 + 5$
- **d** $12.64 + 2.32 \times 1.3$
- **e** $6.5 \times 1.3 - 5.06$
- **f** $(6.7 \div 8) + 1.6$
- **g** $1.453 + \dfrac{7.6}{3.2}$
- **h** $\dfrac{5.34 + 3.315}{4.03}$
- **i** $\dfrac{6.54}{2.3} - 1.08$
- **j** $\dfrac{5.27}{1.4 \times 1.35}$
- **k** $\dfrac{11.5}{2.9 - 1.43}$
- **l** $\dfrac{0.23 \times 4.26}{1.32 + 3.43}$
- **m** $8.9 - \dfrac{8.9}{10.4}$
- **n** $\dfrac{12.6}{8.3} - \dfrac{1.98}{4.62}$
- **o** $12.9 - 2.03^2$
- **p** $(9.4 - 2.67)^3$
- **q** $12.02^2 - 7.05^2$
- **r** $\left(\dfrac{16.8}{9.3} - 1.01\right)^2$
- **s** $\dfrac{4.07^2}{8.2 - 4.09}$
- **t** $6.8 + \dfrac{1.4}{6.9} - \dfrac{1.2}{9.3}$
- **u** $4.3 + \left(1.2 + \dfrac{1.6}{5}\right)^2$
- **v** $\dfrac{6.1}{2.8} + \left(\dfrac{2.1}{1.6}\right)^2$
- **w** $6.4 - (1.2^2 + 1.9^2)^2$
- **x** $\left(4.8 - \dfrac{1}{9.6}\right) \times 4.3$

TIP

Remember the order of operations using BODMAS:

Brackets
Orders
Divide
Multiply
Add
Subtract

Some people remember the order of operations as BIDMAS – I stands for indices.

1.7 Rounding and estimating

KEY LEARNING STATEMENTS

- You may be asked to round numbers to a given number of decimal places or to a given number of significant figures.
- To round to a decimal place:
  - look at the value of the digit to the right of the place you are rounding to
  - if this value is ⩾5 then you round up (add 1 to the digit you are rounding to)
  - if this value is ⩽4 then leave the digit you are rounding to as it is.
- To round to a significant figure:
  - the first *non*-zero digit (before or after the decimal place in a number) is the first significant figure
  - find the correct digit and then round off from that digit using the rules above.
- Estimating involves rounding values in a calculation to numbers that are easy to work with (usually without the need for a calculator).
- An estimate allows you to check that your calculations make sense.

KEY CONCEPTS

- Rounding numbers to a given number of decimal places or significant figures.
- Estimating an approximate answer.

1 Round these numbers to:
  i   two decimal places
  ii one decimal place
  iii the nearest whole number.

  a 5.6543   b 9.8774   c 12.8706   d 0.009
  e 10.099   f 45.439   g 13.999   h 26.001

2 Round each of these numbers to three significant figures.

  a 53 217   b 712 984   c 17.364   d 0.007279

3 Round the following numbers to two significant figures.

  a 35.8   b 5.234   c 12 345   d 0.00875
  e 432 128   f 120.09   g 0.00456   h 10.002

4 Use whole numbers to show why these estimates are correct.

  a 3.9×5.1 is approximately equal to 20
  b 68×5.03 is approximately equal to 350
  c 999×6.9 is approximately equal to 7000
  d $42.02 \div 5.96$ is approximately equal to 7

5 Estimate the answers to each of these calculations to the nearest whole number.

  a $5.2 + 16.9 - 8.9 + 7.1$
  b $(23.86 + 9.07) \div (15.99 - 4.59)$
  c $\dfrac{9.3 \times 7.6}{5.9 \times 0.95}$
  d $8.9^2 \times \sqrt{8.98}$

TIP

If you are told what degree of accuracy to use, it is important to round to that degree. If you are not told, you can round to three significant figures.

REVIEW EXERCISE

1. List the integers in the following set of numbers.

   $$\frac{3}{4} \quad 24 \quad 0.65 \quad -12 \quad 3\frac{1}{2} \quad 0 \quad -15 \quad 0.66 \quad -17$$

2. List the first five multiples of 15.

3. Find the lowest common multiple of 12 and 15.

4. Write each number as a product of its prime factors.

   a 196 b 1845 c 8820

5. Find the HCF of 28 and 42.

6. Simplify:

   a $\sqrt{100} \div \sqrt{4}$ b $\sqrt{100 \div 4}$ c $(\sqrt[3]{64})^3$ d $4^3 + 9^2$

   e $23 \times \sqrt[4]{1296}$ f $-24 \times \sqrt[3]{343}$ g $\left(\frac{1}{2}\right)^{-2} + \sqrt[5]{1}$ h $\left(\frac{1}{2}\right)^{-4} - \sqrt[6]{46656}$

7. Calculate. Give your answer correct to two decimal places.

   a $\dfrac{5.4 \times 12.2}{4.1}$ b $\dfrac{12.2^2}{3.9^2}$ c $\dfrac{12.65}{2.04} + 1.7 \times 4.3$

   d $\dfrac{3.8 \times 12.6}{4.35}$ e $\dfrac{2.8 \times 4.2^2}{3.3^2 \times 6.2^2}$ f $2.5 - \left(3.1 + \dfrac{0.5}{5}\right)^2$

8. Write each of the following in the form of 3^x.

   a 1 b 27 c $\dfrac{1}{9}$ d $\dfrac{1}{3}$

   e $3^4 \times 3^{-2}$ f $\dfrac{3^8}{3^8}$ g $(3^2)^4$ h $(3^{-2})^2$

9. Simplify. Leave your answers in index notation.

   a $\dfrac{3^4 \times 3^7}{3^4}$ b $\dfrac{2^5 \times 2^4}{2^3}$ c $\dfrac{2^3 \times 2^{-4}}{2^2 \times 2^{-2}}$ d $\dfrac{4 \times 4^{-3}}{4^{-2} \times 4^0}$

10. Round each number to three significant figures.

    a 1235.6 b 0.76513 c 0.0237548 d 31.4596

11. Naresh has 6400 square tiles. Is it possible for him to arrange these to make a perfect square? Justify your answer.

12. Ziggy has a square sheet of fabric with sides 120 cm long. Is this big enough to cover a square table of area 1.4 m²? Explain your answer.

13. A cube has a volume of 3.375 m³. How high is it?

14. Estimate the answer to each of these calculations to the nearest whole number.

    a 9.75×4.108 b $0.0387 \div 0.00732$

    c $\dfrac{36.4 \times 6.32}{9.987}$ d $\sqrt{64.25} \times 3.098^2$

TIP

If there are brackets, fractions or roots in your calculation you need to enter these correctly on the calculator. When there is more than one term in the denominator, the calculator will divide by the first term only unless you enter brackets.

Chapter 2: Making sense of algebra

2.1 Using letters to represent unknown values

KEY LEARNING STATEMENTS

- Letters in algebra are called variables because they can have many different values (the value varies). Any letter can be used as a variable, but x and y are used most often.
- A number on its own is called a constant.
- A term is a group of numbers and/or variables combined by the operations multiplying and/or dividing only.
- An algebraic expression links terms by using the + and − operation signs. An expression does not have an equals sign (unlike an equation). An expression can have just one term.

KEY CONCEPTS

- Using letters to represent unknown values.
- Writing expressions to represent mathematical information.

1 Write expressions, in terms of x, to represent:

    a   a number times seven

    b   the sum of a number and 12

    c   five times a number minus two

    d   the difference between a third of a number and twice the number.

2 Sal is p years old.

    a   How old will Sal be in five years' time?

    b   How old was Sal four years ago?

    c   Li is four times Sal's age. How old is Li?

3 Three people win a prize of $\$x$.

    a   If they share the prize equally, how much will each of them receive?

    b   If the prize is divided so that the first person gets half as much money as the second person and the third person gets three times as much as the second person, how much will each receive?

TIP

An expression in terms of x means that the variable letter used in the expression is x.

TIP

Being able to translate a word problem into an expression is a useful strategy for problem solving. Remember you can use any letter as a variable as long as you say what it means.

2 Making sense of algebra

2.2 Substitution

KEY LEARNING STATEMENTS

- Substitution involves replacing variables with given numbers to work out the value of an expression. For example, you may be told to evaluate $5x$ when $x = -2$. To do this you work out $5 \times (-2) = -10$

KEY CONCEPT

Substituting numbers for letters and words to find the value of expressions and formulae.

TIP

Remember that the order of operations rules always apply in these calculations.

1 Evaluate the following expressions when $x = 5$.

   a $4x$ b $12x$ c $3x - 4$ d x^2

   e $-2x^2$ f $14 - x$ g $x^3 - 10x$ h $x^3 - x^2$

   i $3(x - 2)$ j $\dfrac{6x}{2}$ k $\dfrac{4x}{10}$ l $\dfrac{80}{x}$

   m $\dfrac{12x}{4}$ n $\dfrac{2x - 4}{2}$ o $\sqrt{9x^2}$ p $\dfrac{3x^3}{2x^2}$

2 Given that $a = 2$, $b = 5$ and $c = -1$, evaluate:

   a abc b $2bc$ c $\dfrac{b^2 + c}{a}$ d $4ac - 3b$

   e $6c - 2ab$ f $2(ab - 4c)$ g $(abc)^3$ h $2(a^2b)^3$

TIP

Take special care when substituting negative numbers. If you replace x with -3 in the expression $4x$, you will obtain $4 \times -3 = -12$, but in the expression $-4x$, you will obtain $-4 \times -3 = 12$.

3 The formula for finding the area (A) of a triangle is $A = \dfrac{1}{2}bh$, where b is the length of the base and h is the perpendicular height of the triangle.
Find the area of a triangle if:

   a   the base is 12 cm and the height is 9 cm

   b   the base is 2.5 metres and the height is 1.5 metres

   c   the base is 21 cm and the height is half as long as the base

   d   the height is 2 cm and the base is the cube of the height.

2.3 Simplifying expressions

KEY LEARNING STATEMENTS

- To simplify an expression you add or subtract like terms.
- Like terms are those that have exactly the same variables (including powers of variables).
- You can also multiply and divide to simplify expressions. Both like and unlike terms can be multiplied or divided.

KEY CONCEPT

Simplifying expressions using the four basic operations.

1 Simplify the following expressions.

- **a** $5m + 6n - 3m$
- **b** $5x + 4 + x - 2$
- **c** $a^2 + 4a + 2a - 5$
- **d** $y^2 - 4y - y - 2$
- **e** $3x^2 + 6x - 8x + 3$
- **f** $x^2y + 3x^2y - 2yx$
- **g** $2ab - 4ac + 3ba$
- **h** $x^2 + 2x - 4 + 3x^2 - y + 3x - 1$

TIP

Remember, like terms must have exactly the same variables with exactly the same indices. So $3x$ and $2x$ are like terms but $3x^2$ and $2x$ are not like terms.

TIP

Remember that although it is better to write variables in alphabetical order, multiplication can be done in any order, so $ab = ba$. This means you can simplify $3ab + 2ba$ to $5ab$.

2 Simplify.

- **a** $4x \times 3y$
- **b** $4a \times 2b$
- **c** $x \times x$
- **d** $3 \times -2x$
- **e** $-6m \times 5n$
- **f** $3xy \times 2x$
- **g** $-2xy \times -3y^2$
- **h** $-2xy \times 2x^2$
- **i** $12ab \div 3a$
- **j** $12x \div 48xy$
- **k** $\dfrac{33abc}{11ca}$
- **l** $\dfrac{45mn}{20n}$
- **m** $\dfrac{80xy^2}{12x^2y}$
- **n** $\dfrac{-36x^3}{-12xy}$
- **o** $\dfrac{y}{x} \times \dfrac{2y}{x}$
- **p** $\dfrac{xy}{2} \times \dfrac{y}{x}$
- **q** $5a \times \dfrac{3a}{4}$
- **r** $7 \times \dfrac{-2y}{5}$
- **s** $\dfrac{x}{4} \times \dfrac{2}{3y}$
- **t** $\dfrac{3x}{5} \times \dfrac{9x}{2}$

TIP

Remember,
$x \times x = x^2$
$y \times y \times y = y^3$
$x \div x = 1$

2.4 Working with brackets

KEY LEARNING STATEMENTS

- You can remove brackets from an expression by multiplying everything inside the brackets by the value (or values) outside the bracket.
- Removing brackets is also called expanding the expression.
- When you remove brackets in part of an expression you may end up with like terms. Add or subtract any like terms to simplify the expression fully.
- In general terms $a(b + c) = ab + ac$.

KEY CONCEPT

Using brackets and expanding products of algebraic expressions.

1 Expand.

   a $3(x + 2)$
   b $2(x - 4)$
   c $-2(x + 3)$
   d $-3(3 - 2x)$
   e $x(x + 3)$
   f $x(2 - x)$
   g $-x(2 + 2x)$
   h $3x(x - 3)$
   i $-2x(2 - 5x)$
   j $-(x - 2)$
   k $-2x(2y - 2x)$
   l $-x(2x - 4)$

TIP

Remember the rules for multiplying integers:

$+ \times + = +$

$- \times - = +$

$+ \times - = -$

If the quantity in front of a bracket is negative, the signs of the terms inside the bracket will change when the brackets are expanded.

2 Remove the brackets and simplify where possible.

   a $2x(x - 2)$
   b $(y - 3)x$
   c $(x - 2) - 3x$
   d $-2x - (x - 2)$
   e $(x - 3)(-2x)$
   f $2(x + 1) - (1 - x)$
   g $x(x^2 - 2x - 1)$
   h $-x(1 - x) + 2(x + 3) - 4$

3 Remove the brackets and simplify where possible.

   a $2x\left(\frac{1}{2}x + \frac{1}{4}\right)$
   b $-3x(x - y) - 2x(y - 2x)$
   c $-2(4x^2 - 2x - 1)x$
   d $(x + y) - \left(\frac{1}{2}x - \frac{1}{2}y\right)$
   e $2x(2x - 2) - x(x + 2)$
   f $x(1 - x) + x(2x - 5) - 2x(1 + 3x)$

REFLECTION

What are some of the common errors that students might make when they expand expressions?

What can you do to avoid making these types of errors?

2.5 Indices

KEY LEARNING STATEMENTS

- An index (also called a power or exponent) shows how many times the base is multiplied by itself.
- x^2 means $x \times x$ and $(3y)^4$ means $3y \times 3y \times 3y \times 3y$.
- The laws of indices are used to simplify algebraic terms and expressions.
- When you are asked to simplify an expression that contains negative indices you apply the same laws as for other indices.
- If you have an expression like $2^x = 16$, you can write the number 16 as a power of 2 to work out the value of x because when $a^x = a^y$, $x = y$.

KEY CONCEPTS

- Positive, negative and zero indices.
- Using the rules of indices in algebra.

1 Simplify.

a $x^9 \times x^2$
b $y^{10} \times y^3$
c $2x \times 3x^2$
d $-2x^2 \times -3x^6$
e $x^2y^3 \times x^3y$
f $-2x \times 8x \times -3x^2$
g $(2x^2y)(xy)$
h $-3x^4 \times 9x^8$

2 Simplify.

a $2x^5 \div 3x^3$
b $18x^3 yz^2 \div 6xyz^2$
c $12xy^3 \div 18xy^2$
d $-6x \div -12x^2$
e $21x^2 y \div 14x^4 y^6$
f $\dfrac{12x^3yz^2}{6xy^4z}$
g $\dfrac{14x^2}{2x^3}$
h $\dfrac{16x^2y}{4xy^2}$
i $\dfrac{x^2y}{3xy^2}$
j $\dfrac{x^7y^3}{x^4y^5}$
k $\dfrac{36x^2yz^4}{-24xyz}$
l $\dfrac{9x^3y^{-2}}{18x^{-2}y^4}$

TIP

It is helpful to memorise the laws of indices:

1 $x^m \times x^n = x^{m+n}$
2 $x^m \div x^n = x^{m-n}$
3 $(x^m)^n = x^{mn}$
4 $x^0 = 1$
5 $x^{-m} = \dfrac{1}{x^m}$

TIP

Remember, a fraction is the top value divided by the bottom value. This means $\dfrac{x^m}{x^n} = x^m \div x^n = x^{m-n}$

TIP

Apply the laws of indices and work in this order:

- simplify any terms in brackets
- apply the multiplication law ($x^m \times x^n = x^{m+n}$) to numerators and then to denominators
- cancel numbers if you can
- apply the division law ($x^m \div x^n = x^{m-n}$) if the same letter appears in the numerator and denominator
- express your answer using positive indices.

2 Making sense of algebra

3   Rewrite each of the following using positive indices only.

   a 3^{-2}
   b $3x^{-3}$
   c $xy^{-1} \div 2$
   d $(xy)^{-1}$
   e $(8xy)^{-2}$
   f $\dfrac{1}{(4xy)^{-2}}$
   g $y^{-5} \times y^6$
   h $x^3 y^{-1} \times y^{-3}$
   i $x^3 y \times x^{-1} y^{-3}$
   j $y^6 (x^3)^{-4} \times (x^3 y^{-2})^2$
   k $(3xy^3)^{-2} \times (2x^3 y)^3$
   l $\dfrac{4y^{-2}}{7x^{-3}}$

   TIP

   You can write simplified expressions with negative indices, such as $5x^{-4}$. If, however, the question states positive indices only, you can use the law $x^{-m} = \dfrac{1}{x^m}$ so that $5x^{-4} = \dfrac{5}{x^4}$.

4   Simplify.

   a $(x^3)^2$
   b $(-2x^3)^3$
   c $\left(\dfrac{2x^2}{x}\right)^4$
   d $(x^9)^3$
   e $(-xy^2)^9$
   f $(x^3 y^2)^4$
   g $-2(xy)^3$
   h $2x^2 (2x)^3$
   i $\dfrac{(xy^2)^3}{x^3 y^6}$
   j $(xy)^4 (x^4)^3$
   k $(3x^y)^y$
   l $-(2x^2)^3$

5   Simplify.

   a $\dfrac{x^4 y \times y^2 x^6}{x^4 y^5}$
   b $\dfrac{2x^2 y^4 \times 3x^3 y}{2xy^4}$
   c $\dfrac{2x^5 y^4 \times 2xy^3}{2x^2 y^5 \times 3x^2 y^3}$
   d $\dfrac{x^3 y^7}{xy^4} \times \dfrac{x^2 y^8}{x^3 y}$
   e $\dfrac{2x^7 y^2}{4x^3 y^7} \times \dfrac{10x^8 y^4}{2x^3 y^2}$
   f $\dfrac{x^9 y^6}{x^4 y^2} \div \dfrac{x^3 y^2}{x^5 y}$
   g $\dfrac{10x^5 y^2}{9x^6 y^6} \div \dfrac{3x^3 y}{5x^7 y^4}$
   h $\dfrac{7y^3 x^2}{5y^5 x^4} \div \dfrac{5x^6 y^2}{7x^5 y^3}$
   i $\dfrac{(x^5 y)^2 \times (x^3 y^4)^2}{(x^3 y^3)^3}$
   j $\dfrac{(2x^4 y^2)^3}{(y^3 x^2)^3} \times \dfrac{(x^4 y^4)^2}{3(x^2 y)^2}$
   k $\left(\dfrac{x^2}{y^4}\right)^3 \times \left(\dfrac{x^5}{y^2}\right)^2$
   l $\dfrac{(5x^3 y^2)^3}{4x^7 y^6} \div \left(\dfrac{2xy^3}{5x^2 y^4}\right)^2$

6   Simplify each expression and give your answer using positive indices only.

   a $\dfrac{x^5 y^{-4}}{x^{-3} y^{-2}}$
   b $\dfrac{x^{-4} y^3}{x^2 y^{-1}} \times \dfrac{x^7 y^{-5}}{x^{-4} y^3}$
   c $\dfrac{(2x^{-3} y^{-1})^3}{(y^2 x^{-2})^2}$
   d $\left(\dfrac{x}{y^3}\right)^{-1} \div \dfrac{(x^2)^4}{y^{-3}}$
   e $\dfrac{x^{-10}}{(y^{-4})^2} \div \left(\dfrac{y^2}{x^3}\right)^{-4}$
   f $\left(\dfrac{x^4 y^{-1}}{x^5 y^{-3}}\right)^2 \times \dfrac{(x^{-2} y^6)^2}{2(xy^3)^{-2}}$

7   Work out the value of x in each equation.

   a $6^x = 216$
   b $3^x = 243$
   c $3^x = 81$
   d $5^x = 625$
   e $2^x = \dfrac{1}{32}$
   f $4^x = \dfrac{1}{64}$

SELF ASSESSMENT

Before you work through the review exercise, check your answers for each exercise in this chapter and answer these questions about your work.

- How would you describe your understanding of the work in this chapter?
- What did you do well?
- Is there anything you need to improve? If so, what steps will you take to make the improvements?

CAMBRIDGE IGCSE™ MATHEMATICS: CORE PRACTICE BOOK

REVIEW EXERCISE

1. Write each of the following as an algebraic expression. Use x to represent 'a number'.

    a. A number increased by 12.

    b. A number decreased by four.

    c. Five times a number.

    d. A number divided by three.

    e. The product of a number and four.

    f. A quarter of a number.

    g. A number subtracted from 12.

    h. The difference between a number and its cube.

2. Determine the value of $x^2 - 5x$ if:

    a. $x = 2$
    b. $x = -3$
    c. $x = \dfrac{1}{3}$

3. Evaluate each expression if $a = -1$, $b = 2$ and $c = 0$.

    a. $\dfrac{-2a + 3b}{2ab}$
    b. $\dfrac{b(c - a)}{b - a}$
    c. $\dfrac{a - b^2}{c - a^2}$
    d. $\dfrac{3 - 2(a - 1)}{c - a(b - 1)}$
    e. $a^3b^2 - 2a^2 + a^4b - ac^3$

4.
    a. $M = 9ab$. Find M when $a = 7$ and $b = 10$.

    b. $V - rs = 2uw$. Find V when $r = 8$, $s = 4$, $u = 6$ and $w = 1$.

    c. $\dfrac{V}{30} = \sqrt{h}$. Find V when $h = 25$.

    d. $P - y = x^2$. Find P when $x = 2$ and $y = 8$.

5. Remove the brackets and simplify as fully as possible.

    a. $2(y + 5)$
    b. $4(y - 1)$
    c. $4(3x - 2y)$
    d. $x(y + 2)$
    e. $2(10x - 7y + 3z)$
    f. $2x(3x + 1)$
    g. $2(x + 3) + 1$
    h. $6(x + 3) - 2x$
    i. $3x + 2(6x - 3y)$

6. Simplify each of the following expressions as fully as possible.

    a. $3a + 4b + 6a - 3b$
    b. $x^2 + 4x - x - 2$
    c. $-2a^2b(2a^2 - 3b^2)$
    d. $2x(x - 3) - (x - 4) - 2x^2$
    e. $16x^2y \div 4y^2x$
    f. $10x^2 - \dfrac{5xy}{2x}$

CONTINUED

7 Expand and simplify if possible.

    a $2(4x - 3) + 3(x + 1)$ **b** $3x(2x + 3) - 2(4 - 3x)$

    c $x(x + 2) + 3x - 3(x^2 - 4)$ **d** $x^2(x + 3) - 2x^3 - (x - 5)$

8 Simplify. Give all answers with positive indices only.

    a $\dfrac{15x^7}{18x^2}$ **b** $5x^2 \times \dfrac{3x^5}{x^7}$ **c** $\dfrac{(x^3)^4}{(x^2)^8}$

    d $(2xy^2)^4$ **e** $\left(\dfrac{4x^3}{y^5}\right)^3$ **f** $(x^3y)^2 \times \dfrac{(x^2y^4)^3}{(xy^2)^3}$

    g $(2xy^3)^{-2} \times (3x^2y)^3$ **h** $\dfrac{(x^{-3}y^2)^4}{2(xy^2)^{-3}} \div \left(\dfrac{x^{-3}y^3}{x^2y^{-1}}\right)^2$

Chapter 3: Lines, angles and shapes

3.1 Lines and angles

KEY LEARNING STATEMENTS

- Angles can be classified according to their size:
  - acute angles are <90°
  - right angles are 90°
  - obtuse angles are >90° but <180°
  - reflex angles are >180° but <360°.
- Two angles that add up to 90° are called complementary angles. Two angles that add up to 180° are called supplementary angles.
- The sum of angles on a straight line is 180°.
- The sum of the angles around a point is 360°.
- When two lines intersect (cross), two pairs of vertically opposite angles are formed. Vertically opposite angles are equal.
- When two parallel lines are cut by a transversal, alternate angles are equal, corresponding angles are equal and co-interior angles add up to 180°.
- When alternate or corresponding angles are equal, or when co-interior angles add up to 180° the lines are parallel.

KEY CONCEPTS

- The mathematical vocabulary used to talk about points, lines, angles and shapes.
- How to classify, measure and construct angles.
- Angle relationships and how to use them to calculate unknown angles.

1 Estimate the size of each angle and say what type of angle it is. Then measure each angle with a protractor and give its size in degrees.

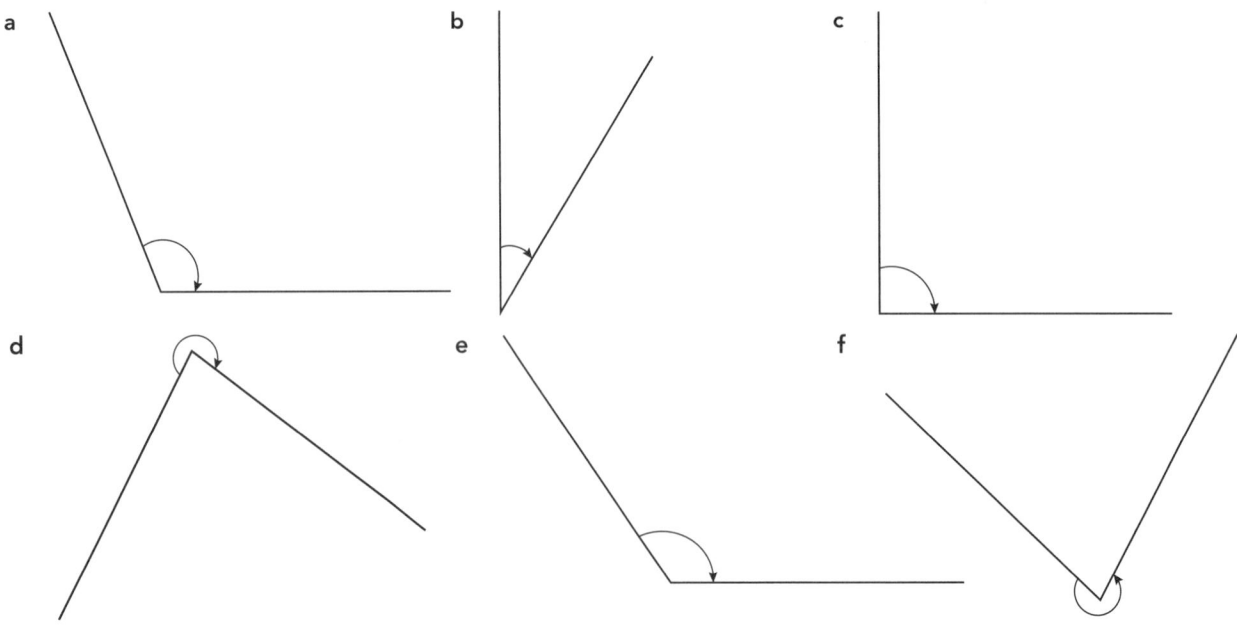

3 Lines, angles and shapes

2 Look at the clock face. Calculate the following.

   a The angle between the hands of the clock at:

      i   3.00 a.m.   ii   18 00 hours.

   b Through how many degrees does the hour hand move between 4.00 p.m. and 5.00 p.m.?

   c Through how many degrees does the minute hand turn in one hour?

   d A clock shows 12 noon. What will the time be when the minute hand has moved 270° clockwise?

3 Will doubling an acute angle always produce an obtuse angle? Explain your answer.

4 Will halving an obtuse angle always produce an acute angle? Explain your answer.

5 What is the complement of each the following angles?

   a   45°   b   62°   c $x°$ d $(90 - x)°$

6 What is the supplement of each of the following angles?

   a   45°   b   90°   c   104°   d $x°$

   e $(180 - x)°$ f $(90 - x)°$ g $(90 + x)°$ h $(2x - 40)°$

7 In the following diagram, PQ and RS are straight lines.

Calculate the sizes of angles x, y and z.
Give reasons for your answers.

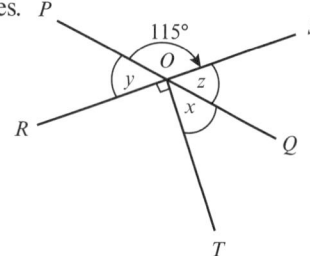

TIP

When you are asked to calculate, it means you must not measure the angle from the diagram.

You need to be able to use the relationships between lines and angles to calculate the values of unknown angles.

8 In the following diagram, MN and PQ are straight lines. Find the size of angle a, giving reasons.

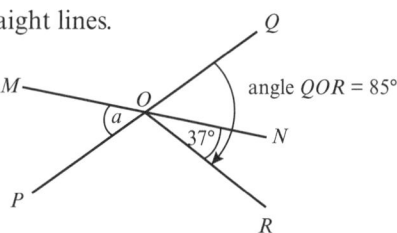

angle QOR = 85°

TIP

Remember, give reasons for statements. Use these abbreviations:

Comp angles
Supp angles
Angles on line
Angles at point
VO angles
Alt angles
Corr angles
Co-int angles

9 Calculate the value of x in each of the following figures.
Give reasons for your answers.

   a 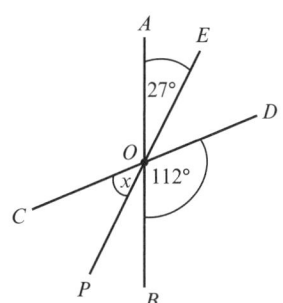   b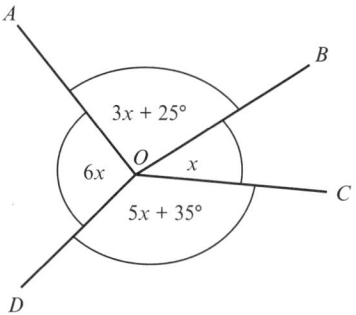

10 In this figure the size of angle *AGH* is given. Calculate the size of all the other angles giving reasons.

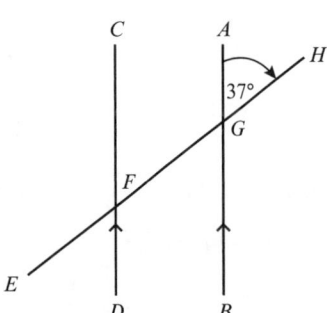

11 Find the values of the angles marked with the letters *x*, *y* and *z* in each diagram. Give reasons for any statements you make.

a
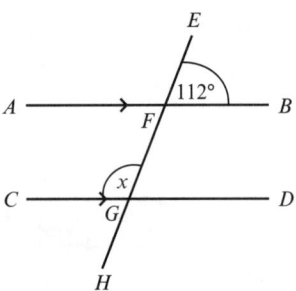

b
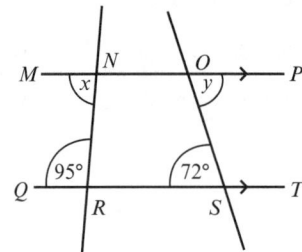

c
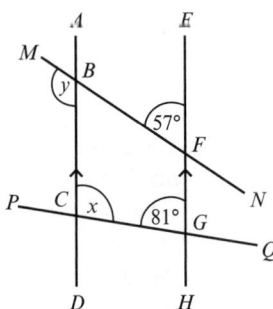

d
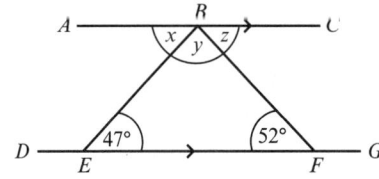

e
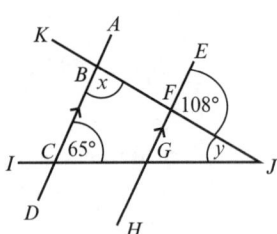

f
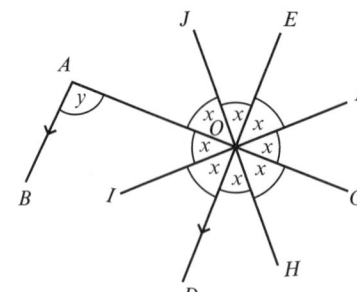

3 Lines, angles and shapes

12 Calculate the value of *x* and *y* in each of the following figures. Give reasons for your answers.

a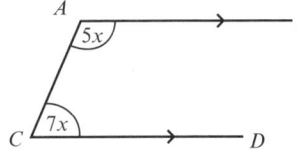

b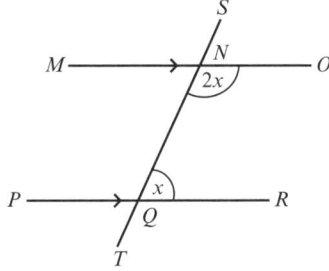

c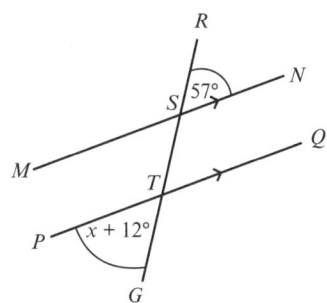

d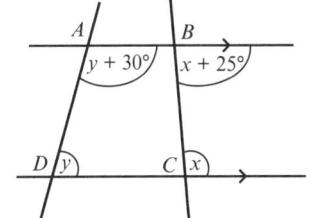

e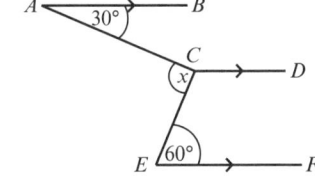

f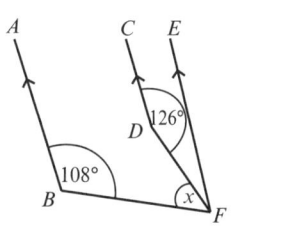

3.2 Triangles

KEY LEARNING STATEMENTS

- Scalene triangles have no equal sides and no equal angles.
- Isosceles triangles have two equal sides. The angles at the bases of the equal sides are equal in size. The converse is also true – if a triangle has two equal angles, then it is an isosceles triangle.
- Equilateral triangles have three equal sides and three equal angles (each being 60°).
- The sum of the interior angles of any triangle is 180°.
- The exterior angle of a triangle is equal to the sum of the two opposite interior angles.

KEY CONCEPTS

- Triangle vocabulary and the properties of triangles.
- Calculating unknown angles using properties of triangles.

1 Find the angles marked with letters. Give reasons for any statements.

a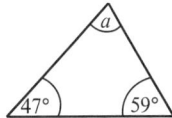

b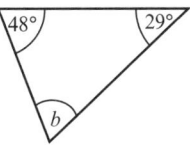

c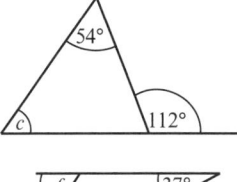

d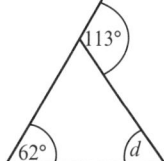

e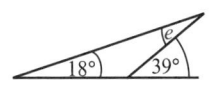

f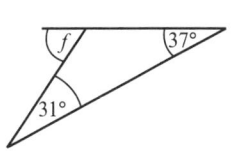

TIP

You may also need to apply the angle relationships for points, lines and parallel lines to find the missing angles in triangles.

21

CAMBRIDGE IGCSE™ MATHEMATICS: CORE PRACTICE BOOK

g 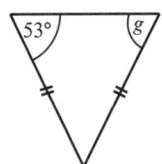   h 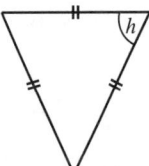   i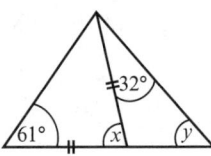

j 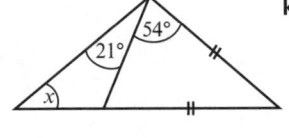   k 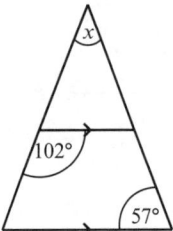   l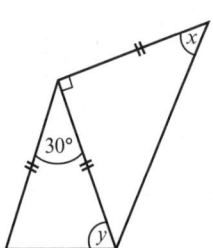

2 Calculate the value of x and hence find the size of the marked angles.

a 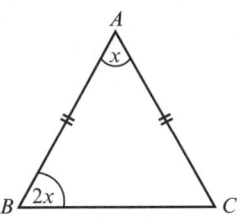   b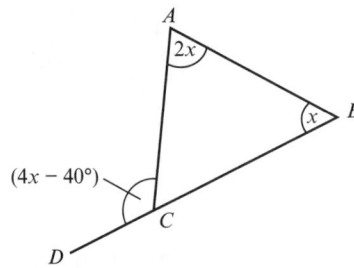

c 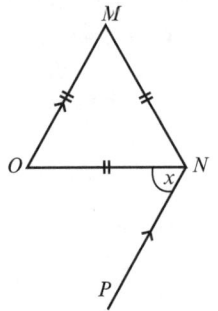   d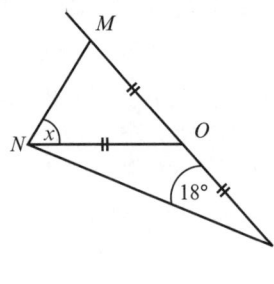

e 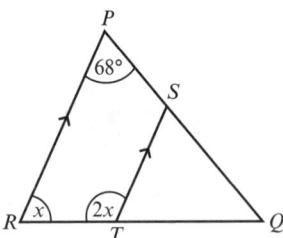   f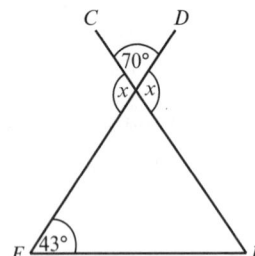

22

3 Lines, angles and shapes

3.3 Quadrilaterals

KEY LEARNING STATEMENTS

- A quadrilateral is a four-sided shape.
  - A trapezium has one pair of parallel sides.
  - A kite has two pairs of adjacent sides equal in length. The diagonals intersect at 90° and the longer diagonal bisects the shorter one. Only one pair of opposite angles is equal. The longer diagonal bisects the angles.
  - A parallelogram has opposite sides equal and parallel. The opposite angles are equal in size and the diagonals bisect each other.
  - A rectangle has opposite sides equal and parallel, and the interior angles all equal 90°. The diagonals are equal in length and they bisect each other.
  - A rhombus is a parallelogram with all four sides equal in length. The diagonals bisect each other at 90° and bisect the angles.
  - A square has four equal sides and four angles each equal to 90°. The opposite sides are parallel. The diagonals are equal in length, they bisect each other at right angles and they bisect the angles.
- The sum of the interior angles of a quadrilateral is 360°.

KEY CONCEPTS

- Quadrilateral vocabulary and the properties of quadrilaterals.
- Calculating unknown angles using the properties of quadrilaterals.

1 Each of the following statements applies to one or more quadrilaterals. For each one, name the quadrilateral(s) to which it always applies.

   a All sides are equal in length.

   b All angles are equal in size.

   c The diagonals are the same length.

   d The diagonals bisect each other.

   e The angles are all 90° and the diagonals bisect each other.

   f Opposite angles are equal in size.

   g The diagonals intersect at right angles.

   h The diagonals bisect the angles.

   i One diagonal divides the quadrilateral into two isosceles triangles.

TIP

The angle relationships for parallel lines will apply when a quadrilateral has parallel sides.

2 Calculate the size of the marked angles in the following diagrams. Give reasons or state the properties you are using.

   a 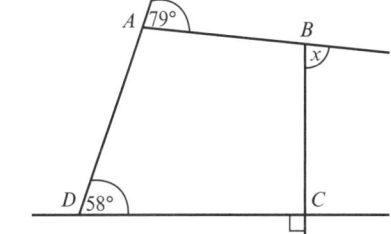   b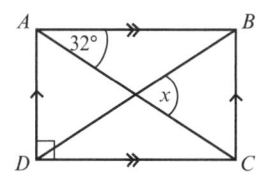

c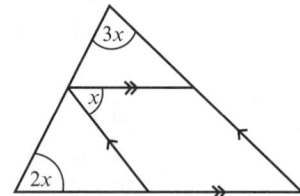

d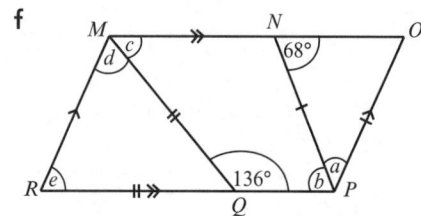

e

f

3 How well do you know the names and properties of quadrilaterals?

Use the flow chart to work out which quadrilateral names could replace the letters A to E.

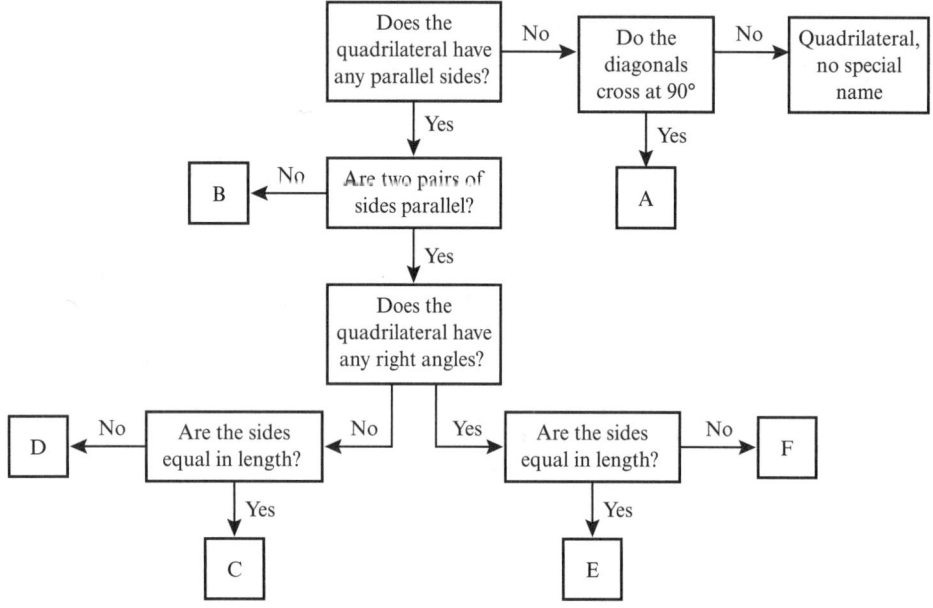

SELF ASSESSMENT

Check your answers to question 3.

Choose the comment that best fits your work.

- Excellent. You know this very well.
- Good. Check the one you got wrong.
- You can improve if you revise this work.

3.4 Polygons

> **KEY LEARNING STATEMENTS**
>
> - A polygon is a two-dimensional shape with three or more sides. Polygons are named according the number of sides they have:
>   - triangle (3)
>   - quadrilateral (4)
>   - pentagon (5)
>   - hexagon (6)
>   - heptagon (7)
>   - octagon (8)
>   - nonagon (9)
>   - decagon (10).
> - A regular polygon has all its sides equal and all its angles equal.
> - The interior angle sum of any polygon can be worked out using the formula $(n - 2) \times 180°$ where n is the number of sides. Once you have the angle sum, you can find the size of one angle of a regular polygon by dividing the total by the number of angles.
> - The sum of the exterior angles of any convex polygon is 360°.

> **KEY CONCEPTS**
>
> - Polygon vocabulary and properties of regular polygons.
> - Calculating unknown angles using the properties of polygons.

> **TIP**
>
> If you can't remember the formula, you can find the size of one interior angle of a regular polygon using the fact that the exterior angles add up to 360°. Divide 360 by the number of angles to find the size of one exterior angle. Then use the fact that the exterior and interior angles form a straight line (180°) to work out the size of the interior angle.

1. For a regular hexagon.

    a Calculate the size of one exterior angle.

    b Find the sum of the interior angles.

    c What is the size of each interior angle?

2. Find the sum of the interior angles of:

    a a regular octagon

    b a regular decagon

    c a regular 15-sided polygon.

3. A coin is made in the shape of a regular heptagon. Calculate the size of each interior angle.

4. The interior angle of a regular polygon is 162°. How many sides does the polygon have?

5. One exterior angle of a regular polygon is 14.4°.

    a What is the size of each interior angle?

    b How many sides does the polygon have?

3.5 Circles

KEY LEARNING STATEMENTS

- A circle is a set of points equidistant from a fixed centre. Half a circle is a semicircle.
- The perimeter of a circle is called its circumference.
- The distance across a circle (through the centre) is called its diameter. A radius is half a diameter.
- An arc is part of the circumference of a circle.
- A chord is a line joining two points on the circumference. A chord cuts the circle into two segments.
- A 'slice' of a circle, made by two radii and the arc between them on the circumference, is called a sector.
- A tangent is a line that touches a circle at only one point.

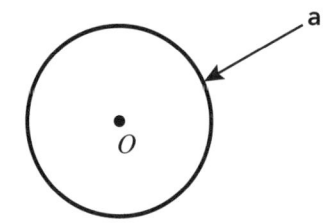

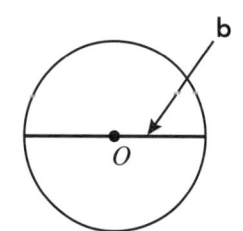

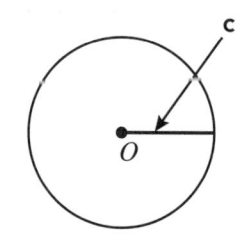

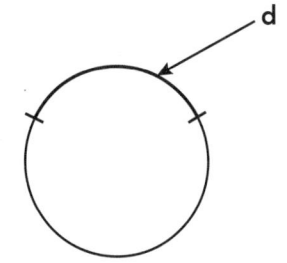

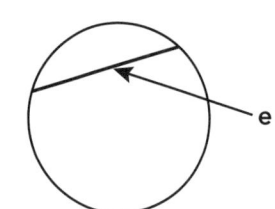

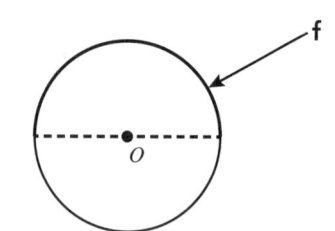

KEY CONCEPT

Circle vocabulary.

Name the parts of each circle shown by the arrows.

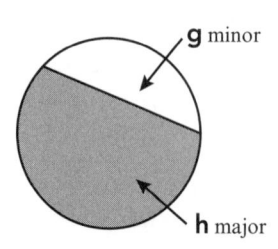

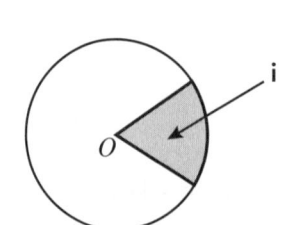

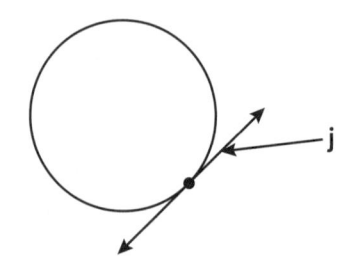

3.6 Construction

> **KEY LEARNING STATEMENTS**
>
> - You need to be able to use a ruler and a pair of compasses to construct triangles given the lengths of three sides.

> **KEY CONCEPT**
>
> How to use a ruler and a pair of compasses to construct a triangle when you know the lengths of the sides.

> **TIP**
>
> Always start with a rough sketch. Label your rough sketch so you know what lengths you need to measure.

1   These steps for constructing a triangle of sides 9 cm, 7 cm and 15 cm using only a ruler and a pair of compasses have been mixed up.

   A   At the endpoint of the line segment draw an arc with a radius of 7 cm.
   B   Draw a line segment 15 cm long.
   C   Draw a line from the intersection to each endpoint.
   D   At the endpoint of the line segment, draw an arc with radius of 9 cm.
   E   Find the point where the two arcs intersect.

   a   Write the letters A to E in a correct order.
   b   Why is there more than one way of ordering them?

2   Construct triangle ABC with $AC = 7$ cm, $CB = 6$ cm and $AB = 8$ cm.

3   Construct triangle MNO with $MN = 4.5$ cm, $NO = 5.5$ cm and $MO = 8$ cm.

4   Construct triangle DEF with $DE = 100$ mm, $FE = 70$ mm and $DF = 50$ mm. What type of triangle is DEF?

5   Nino is trying to construct triangle ABC with sides $AB = 7$ cm, $BC = 3$ cm and $AC = 8$ cm. He started by drawing an 8 cm long line segment and then used a pair of compasses to draw arcs like this:

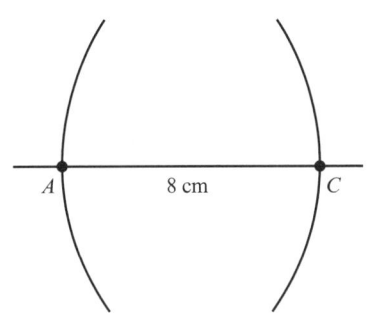

   a   What has he done incorrectly?
   b   Construct the triangle correctly.

> **REFLECTION**
>
> List three things you can do to make your constructions as accurate as possible. How does each thing on your list help?

REVIEW EXERCISE

1 Write a correct mathematical definition for each of the following:

    a alternate angles   **b** an isosceles triangle

    c a kite   **d** a rhombus

    e a regular polygon   **f** an octagon

    g circumference   **h** transversal.

2 Find the value of the marked angles in each of the following.

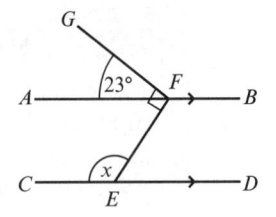

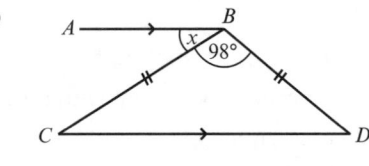

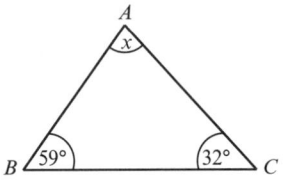

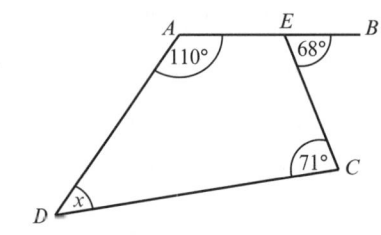

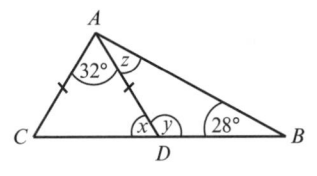

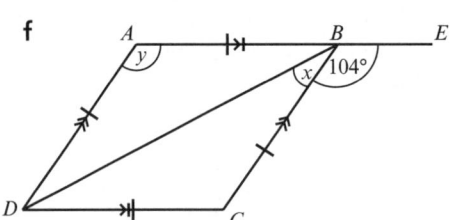

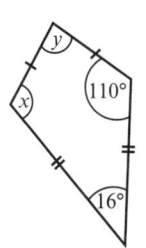

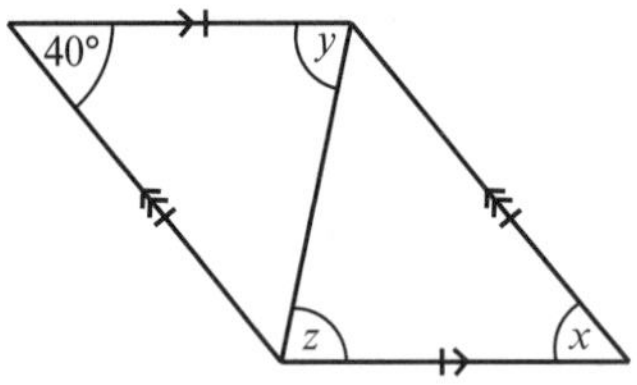

CONTINUED

3 For each shape combination find the size of angle x. All shapes in both diagrams are regular polygons.

a

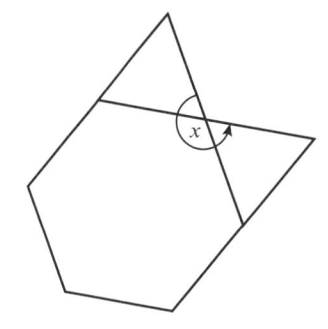

b

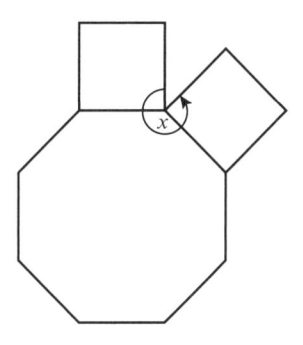

4 Use the diagram of the circle with centre O to answer these questions.

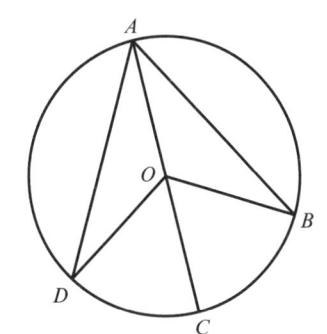

  a What are the correct mathematical names for:

   i DO

   ii AB

   iii AC?

  b Four radii are shown on the diagram. Name them.

  c If OB is 12.4 cm long, how long is AC?

  d Draw a copy of the circle and draw the tangent to the circle that passes through point B.

5 Points A and B are the centres of two overlapping circles of diameter 9 cm.

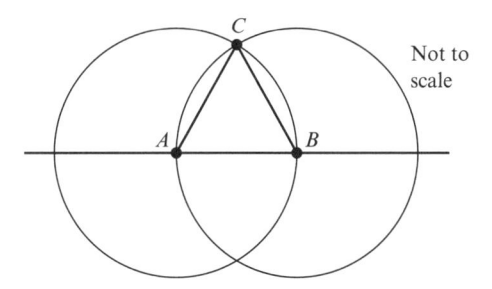

Not to scale

Accurately construct the triangle without drawing the circles.

Chapter 4: Collecting, organising and displaying data

4.1 Collecting and classifying data

KEY LEARNING STATEMENTS

- Data is a set of facts, numbers or other information, collected to try to answer a question.
- Primary data is 'original' data. You can collect it by measuring, observation, doing experiments, carrying out surveys or asking people to complete questionnaires.
- Secondary data is data from a non-original source. For example, you could find the area of each of the world's oceans by referring to an atlas.
- You can classify data as qualitative or quantitative.
- Qualitative data is non-numeric such as colour, make of vehicle or favourite flavour.
- Quantitative data is numerical data that was counted or measured. For example, age, marks in a test, shoe size, height.
- Quantitative data can be discrete or continuous.
- Discrete data can only take certain values and is usually something counted. For example, the number of children in your family. There are no in-between values; you can't have $2\frac{1}{2}$ children in a family.
- Continuous data can take any value and is usually something measured. For example, the heights of trees in a rainforest could range from 50 to 60 metres. Any value in-between those two heights is possible.

KEY CONCEPT

Collecting, classifying and tabulating data.

The following table of data was collected about ten students in a high school. Study the table and then answer the questions about the data.

Grade	10	10	11	11	11	10	11	10	10	11
Height (metres)	1.55	1.61	1.63	1.60	1.61	1.62	1.64	1.69	1.61	1.65
Shoe size	3	4	7	6	9	7	8	7	5	10
Mass (kg)	40	51	68	54	60	43	70	56	51	55
Eye colour	Br	Gr	Gr	Br	Br	Br	Br	Gr	Bl	Br
Hair colour	Bla	Bla	Blo	Br	Br	Br	Bla	Bla	Bla	Bla
No. of siblings	0	3	4	2	1	2	3	1	0	3

Key:
Br: Brown
Bla: Black
Blo: Blonde
Bl: Blue
Gr: Green

1 Which of these data categories are qualitative?
2 Which of these data categories are quantitative?
3 Which sets of numerical data are discrete data?
4 Which sets of numerical data are continuous data?
5 How do you think each set of data was collected? Give a reason for your answers.

4.2 Organising data

KEY LEARNING STATEMENTS

- Once data has been collected, you need to arrange and organise it so that it is easier to work with, interpret and draw conclusions from.
- You can use tally tables, frequency tables and stem-and-leaf diagrams to organise data and to show the totals of different values or categories. You can use a back-to-back stem-and-leaf diagram to show two sets of related data.
- When you have a large set of numerical data, with lots of different values, you can group the data into intervals called class intervals. Class intervals should not overlap.
- You can use a two-way table to show the frequency of results for two or more sets of data.

KEY CONCEPTS

- Reading, interpreting and drawing conclusions from tables and statistical diagrams.
- Comparing data sets.
- Restrictions on drawing conclusions from given data.
- Constructing and interpreting stem-and-leaf diagrams.

TIP

In data handling, the word frequency means the number of times a score or observation occurs.

1 Here are the marks out of 10 obtained by 40 students in an assignment.

6	5	6	7	4	5	8	6	7	10
7	6	5	6	1	9	4	4	2	6
5	5	7	3	4	5	8	3	5	8
10	9	9	7	5	5	7	6	4	2

Copy and complete this tally table to organise the data.

Mark	Tally	Frequency
1		
2		

2   Nika rolled a dice 40 times and got these results.

6	6	6	5	4	3	2	6	5	4
1	1	3	2	5	4	3	3	3	2
1	6	5	5	4	4	3	2	5	4
6	3	2	4	2	1	2	2	1	5

a   Copy and complete this frequency table to organise the data.

Score	1	2	3	4	5	6
Frequency						

b   Do the results suggest that this is a fair dice or not? Give a reason for your answer.

3   These are the percentage scores of 50 students in an examination.

54	26	60	40	55	82	67	59	57	70
67	44	63	56	46	48	55	63	42	58
45	54	76	65	63	61	49	54	54	53
67	56	69	57	38	57	51	55	59	78
65	52	55	78	69	71	73	88	80	91

a   Copy and complete this grouped frequency table to organise the results.

Score	0–29	30–39	40–49	50–59	60–69	70–79	80–100
Frequency							

b   How many students scored 70% or more?

c   How many students scored lower than 40%?

d   How many students scored 40% or more but less than 60%?

e   The first and last class intervals in the table are greater than the others. Suggest why this is the case.

f   Draw an ordered stem-and-leaf diagram to show the data. What advantage does it have over the frequency table?

4 This is a section of the table you worked with in Exercise 4.1.

Student	1	2	3	4	5	6	7	8	9	10
Eye colour	Br	Gr	Gr	Br	Br	Br	Br	Gr	Bl	Br
Hair colour	Bla	Bla	Blo	Br	Br	Br	Bla	Bla	Bla	Bla

a Copy and complete this two-way table using data from the table.

Eye colour / Hair colour	Brown	Blue	Green
Blonde			
Brown			
Black			

b Write a sentence to summarise what you can conclude from this table.

c Do you think you will find similar patterns in eye and hair colour in your own school? Explain your answer.

5 A traffic department installed a camera and counted the number of cars passing an intersection every hour for 24 hours.

These are the results:

```
 1   5   7  12  16  23  31  31
51  35  31  33  29  24  43  48
41  39  37  20  19  18  12   2
```

a Draw an ordered stem-and-leaf diagram to show the data.

b What is the maximum number of cars that passed through in an hour during this period?

REFLECTION

Look back at questions 1, 2 and 3. What methods did you use to make sure you recorded all the data points correctly and that you included all of the data?

Why is it important to check that you have included all the data correctly?

4.3 Using charts to display data

> **KEY LEARNING STATEMENTS**
>
> - Charts usually help you to see patterns and trends in data more easily than tables.
> - Pictograms use symbols to show the frequency of data in different categories. They are useful for discrete, categorical and ungrouped data.
> - Bar charts are useful for categorical and ungrouped data. A bar chart has bars of equal width which are equally spaced.
> - Bar charts can be drawn as horizontal or vertical charts. They can also show two or more sets of data on the same set of axes.
> - Pie charts are circular graphs that use sectors of a circle to show the proportion of data in each category.

> **KEY CONCEPT**
>
> - Constructing and interpreting bar charts, pie charts and pictograms.
> - Graphs in practical situations.

1 Study the diagram carefully and answer the questions about it.

   a What type of chart is this?

   b What does the chart show?

   c What does each full symbol represent?

   d How are 15 students shown on the chart?

   e How many students are there in Year 8?

   f Which year group has the most students? How many are there in this year group?

   g Do you think these are accurate or rounded figures? Why?

2 Naresh did a survey to find out how much time his friends spent on social media during the day. These are his results.

Person	Alain	Li	Zayn	David
Time (hours)	$2\frac{1}{2}$	$1\frac{1}{3}$	$4\frac{3}{4}$	$2\frac{1}{4}$

Draw a pictogram to show this data.

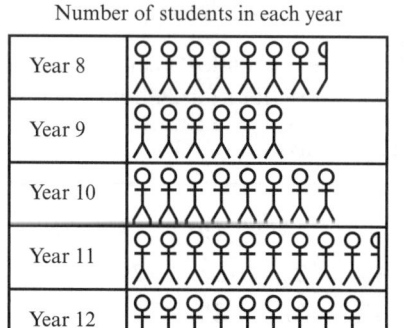

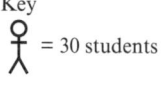

> **TIP**
>
> Choose symbols that are easy to draw and to divide into parts. If it is not given, choose a suitable scale for your symbols so you don't have to draw too many.

3 Study the two bar charts.

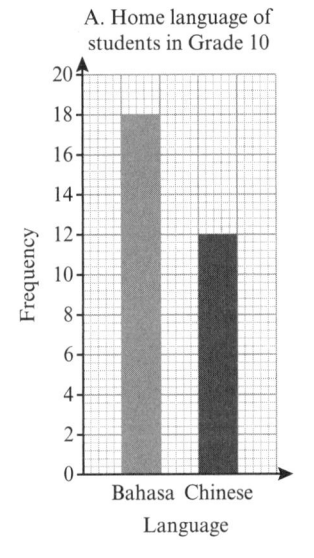

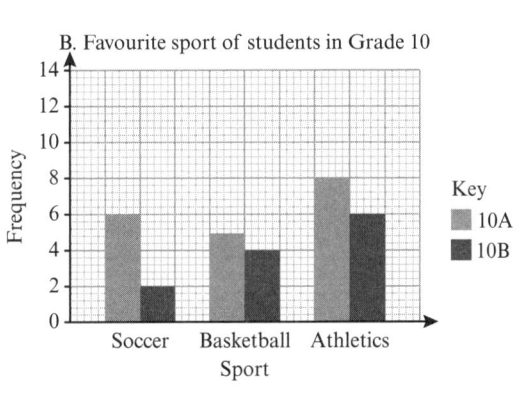

A. Home language of students in Grade 10

B. Favourite sport of students in Grade 10

Key
- 10A
- 10B

> **TIP**
>
> Dual and composite bar charts show two or more sets of data on the same pair of axes. These bar graphs need a key to show what each bar or part of a bar represents.

a What does chart A show?

b How many students have Bahasa as a home language?

c How many students were included in the survey?

d What does chart B show?

e Which sport is most popular in Class 10A?

f Which sport is most popular in Class 10B?

g How many students chose basketball as their favourite sport?

4 The table below shows the type of food students in a hostel chose for breakfast.

	Cereal	Hot porridge	Bread
Grade 10	8	16	12
Grade 11	2	12	10

a Draw a single bar chart to show the total number of students who chose each breakfast food.

b Draw a composite bar chart to show the breakfast food choice for students by grade.

5 Jyoti recorded the number and type of 180 vehicles passing her home in Bengaluru. She drew this pie chart to show her results.

   a Which type of vehicle was most common?

   b What percentage of the vehicles were tuk-tuks?

   c How many trucks passed Jyoti's home?

   d Which types of vehicles were least common?

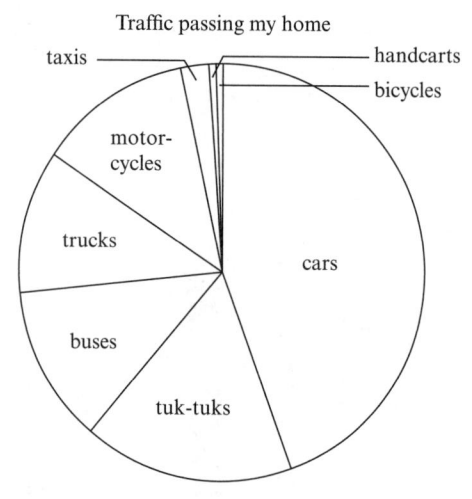

6 In an exam the results for 120 students were: 5% attained an A grade, 12% attained a B grade, 41% attained a C grade, 25% attained a D grade and the rest attained E grade or lower.

   a Represent this information on a pie chart.

   b How many students attained an A?

   c How many students attained a D or lower?

   d Which grade was attained by the most students?

7 The line graphs represent the average monthly temperature and the average monthly rainfall in the desert in Egypt.

   a What is the maximum temperature?

   b In what months is the average temperature above 20 °C?

   c Is Egypt in the northern or southern hemisphere?

   d Is the temperature ever below freezing point?

   e What is the average rainfall in November?

   f In which month is the average rainfall 7 mm?

   g Looking at both graphs, what can you say about the rainfall when the temperatures are highest?

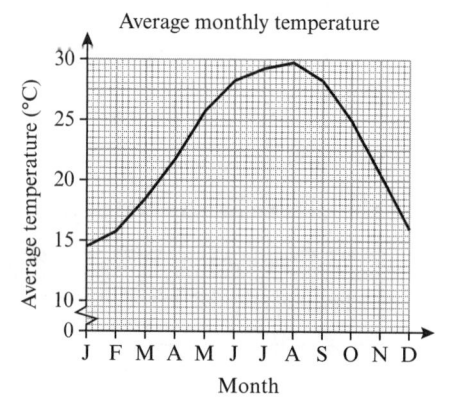

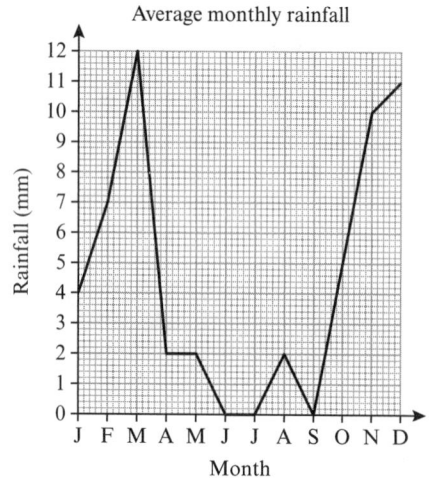

SELF ASSESSMENT

1 Use this checklist to assess your own pie chart. Write the letters A to E in your book and write 'Yes', 'No' or 'Partially complete' next to each.

   Criteria

   A Does the graph have a clear heading?

   B Is the key clear and complete?

   C Are the sectors on the graph clearly labelled?

   D Is each sector the correct size?

   E Are you able to use the graph to answer questions about the data?

2 Write a comment about your work.

REVIEW EXERCISE

1 Mika collected data about how many children different families in her community had. These are her results.

0	3	4	3	3	2	2	2	2	1	1	1
3	3	4	3	6	2	2	2	0	0	2	1
5	4	3	2	4	3	3	3	2	1	1	0
3	1	1	1	1	0	0	0	2	4	5	3

   a How do you think Mika collected the data?

   b Is this data discrete or continuous? Why?

   c Is this data qualitative or quantitative? Why?

   d Draw a frequency table, with tallies, to organise the data.

   e Represent the data on a pie chart.

   f Write a statement to compare the number of families that have three or fewer children with those that have four or more children.

CONTINUED

2 The heights of two groups of students are given in the table to the nearest centimetre.

10A					10B				
183	159	166	165	184	145	161	171	162	154
167	178	175	178	175	164	157	180	166	147
185	174	187	176	166	159	164	163	162	171

   a Draw a back-to-back stem-and-leaf diagram to organise the data.

   b How many students in total are taller than 170 cm?

   c What does the diagram suggest about the heights of each group?

3 Mrs Sanchez bakes and sells cookies. One week she sells 420 peanut crunchies, 488 chocolate cups and 320 coconut munchies. Draw a pictogram to represent this data.

4 Study the chart below.

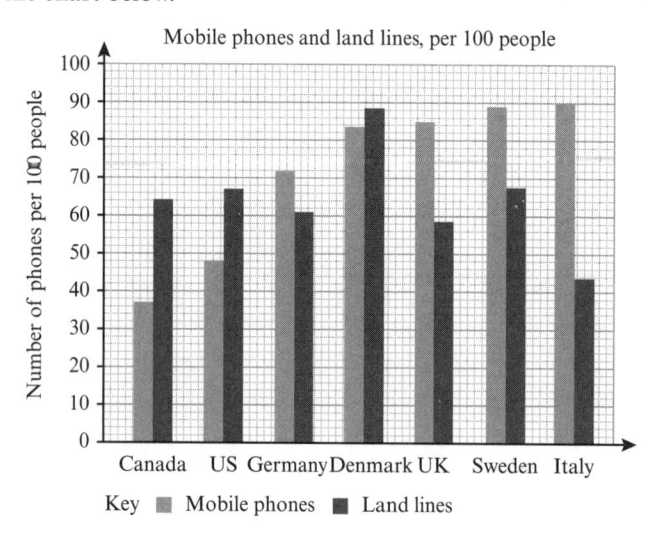

   a What do you call this type of chart?

   b What does the chart show?

   c Can you tell how many people in each country have a mobile phone from this chart? Explain your answer.

   d In which countries do a greater proportion of the people have a land line than a mobile phone?

   e In which countries do more people have mobile phones than land lines?

   f In which country do more than 80% of the population have a land line and a mobile phone?

   g What do you think the bars would look like for your country? Why?

CONTINUED

5 The bar chart shows the category of goods sold and the value of sales for three stores in different places.

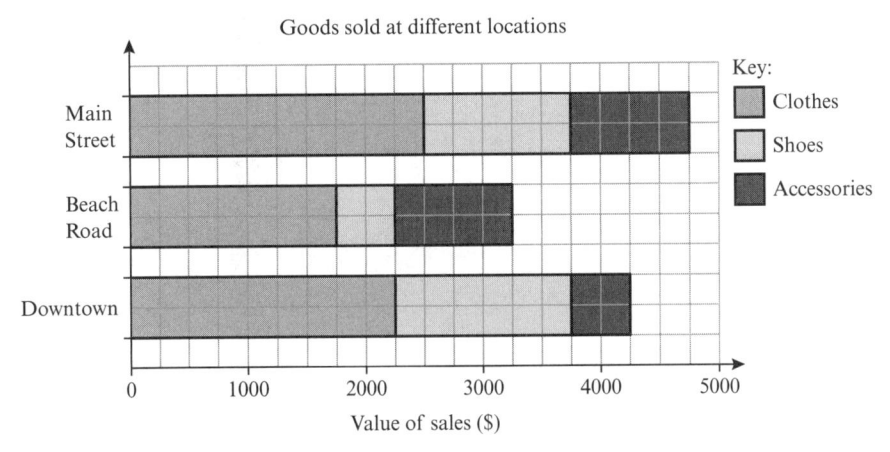

Goods sold at different locations

Key:
- Clothes
- Shoes
- Accessories

Value of sales ($)

a Which store earned most from the sale of shoes?

b What is the total value of sales at the Main Street store?

c What is the total value of accessories sold for all three stores?

d What was the value of Downtown sales excluding accessories?

e What percentage of the total sales value at Beach Road came from selling shoes? Give your answer to the nearest percent.

6 Amy bought a new car in 2022. Its value is shown in the table below, and the line graph represents this information.

Year	Value of car
2022	$13 900
2023	$7000
2024	$5700
2025	$4700
2026	$4000

a Describe the trend shown by the graph.

b Use the graph to estimate the value of the car in 2027.

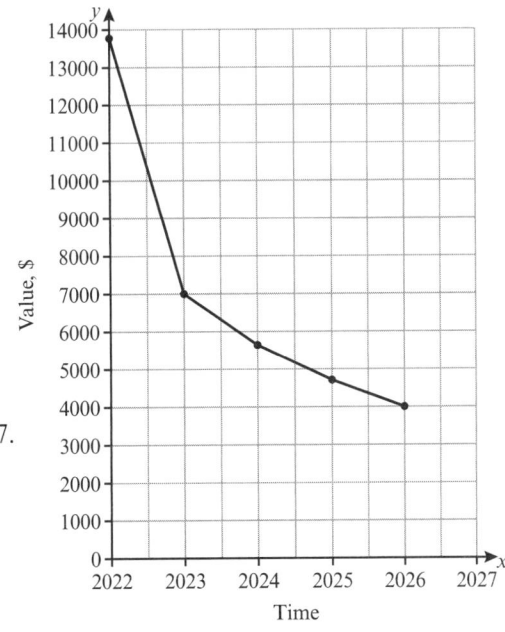

Chapter 5: Fractions, percentages and standard form

5.1 Revisiting fractions

KEY LEARNING STATEMENTS

- Equivalent means, 'has the same value'.
- To find equivalent fractions either multiply both the numerator and denominator by the same number or divide both the numerator and denominator by the same number.

KEY CONCEPTS

- The language and notation of fractions.
- Simplifying and finding equivalent fractions.

TIP

You can cross multiply to make an equation and then solve it. For example:

$$\frac{1}{2} \diagdown \frac{x}{28}$$

$2x = 28$
$x = 14$

1. Write each fraction in its simplest form.

   a $\frac{6}{12}$ b $\frac{4}{12}$ c $\frac{3}{9}$ d $\frac{8}{32}$ e $\frac{12}{48}$

   f $\frac{125}{1000}$ g $\frac{3}{15}$ h $\frac{4}{6}$ i $\frac{24}{32}$ j $\frac{375}{1000}$

2. Find the missing value in each pair of equivalent fractions.

   a $\frac{3}{4} = \frac{x}{44}$ b $\frac{1}{3} = \frac{x}{900}$ c $\frac{1}{2} = \frac{x}{50}$ d $\frac{2}{5} = \frac{26}{x}$ e $\frac{5}{7} = \frac{120}{x}$

   f $\frac{6}{5} = \frac{66}{x}$ g $\frac{11}{9} = \frac{143}{x}$ h $\frac{5}{3} = \frac{80}{x}$ i $\frac{8}{12} = \frac{x}{156}$ j $\frac{7}{9} = \frac{49}{x}$

5.2 Operations on fractions

KEY LEARNING STATEMENTS

- To multiply fractions, multiply numerators by numerators and denominators by denominators. Write mixed numbers as improper fractions before multiplying or dividing.
- To add or subtract fractions change them to equivalent fractions with the same denominator, then add (or subtract) the numerators only.
- To divide one fraction by another, multiply the first fraction by the reciprocal of the second fraction. This means that you invert the second fraction (turn it upside down) and change the ÷ sign to a × sign.
- Give your answers to calculations with fractions in their simplest form. Unless you are told to give your answer in a specific form you can use improper fractions or mixed numbers.

KEY CONCEPT

Calculating with fractions, including the correct order of operations and use of brackets.

5 Fractions, percentages and standard form

1. Rewrite each mixed number as an improper fraction in its simplest form.

   a $2\frac{7}{42}$ b $3\frac{5}{40}$ c $1\frac{12}{22}$ d $9\frac{30}{100}$ e $11\frac{24}{30}$

   f $3\frac{75}{100}$ g $14\frac{3}{4}$ h $2\frac{35}{45}$ i $9\frac{15}{45}$ j $-2\frac{7}{9}$

   TIP
   If you can simplify the fraction part first you will have smaller numbers to multiply to get the improper fraction.

2. Multiply, giving your answers in simplest form.

   a $\frac{1}{5} \times \frac{3}{15}$ b $\frac{1}{4} \times \frac{2}{5}$ c $\frac{2}{3} \times \frac{6}{10}$ d $\frac{3}{5} \times \frac{9}{12}$ e $\frac{2}{11} \times \frac{8}{9}$

   f $\frac{6}{11} \times \frac{2}{3}$ g $\frac{10}{13} \times \frac{3}{7}$ h $\frac{20}{50} \times \frac{9}{15}$ i $\frac{10}{14} \times \frac{3}{4}$ j $\frac{6}{8} \times \frac{3}{11}$

   k $0.6 \times \frac{1.5}{1.8}$ l $0.2 \times \frac{12}{20}$ m $\frac{2.4}{0.6} \times \frac{1.44}{0.4}$ n $\frac{3}{0.5} \times \frac{7}{8}$ o $\frac{1.2}{3.6} \times \frac{0.7}{1.4}$

   TIP
   Remember, you can cancel to simplify when you are multiplying fractions.

3. Calculate.

   a $1\frac{4}{5} \times 12$ b $\frac{9}{13} \times 7$ c $3\frac{1}{2} \times 4$ d $2\frac{1}{3} \times 2\frac{2}{5}$

   e $2 \times 4\frac{1}{2} \times \frac{1}{3}$ f $\frac{1}{5} \times \frac{12}{19} \times 2\frac{1}{2}$ g $\frac{1}{3}$ of 360   h $\frac{3}{4}$ of $\frac{2}{7}$

   i $\frac{8}{9}$ of 81   j $\frac{2}{3}$ of $4\frac{1}{2}$ k $\frac{1}{2}$ of $9\frac{16}{50}$ l $\frac{3}{4}$ of $2\frac{1}{3}$

   TIP
   Remember, the word 'of' tells you to ×.

4. Calculate, giving your answer as a fraction in simplest form.

   a $\frac{3}{4} - \frac{1}{5}$ b $\frac{1}{5} + \frac{1}{6}$ c $\frac{1}{5} - \frac{1}{9}$ d $\frac{1}{6} + \frac{3}{8}$

   e $\frac{2}{3} - \frac{4}{10}$ f $\frac{9}{10} - \frac{7}{12}$ g $\frac{4}{7} + \frac{1}{3}$ h $\frac{2}{3} + \frac{2}{5}$

   i $\frac{7}{8} - \frac{1}{3}$ j $2\frac{1}{2} + 3\frac{1}{3}$ k $2\frac{1}{8} + 1\frac{1}{7}$ l $4\frac{3}{10} + 3\frac{3}{4}$

   m $1\frac{1}{13} - \frac{4}{5}$ n $3\frac{9}{10} - 2\frac{7}{8}$ o $2\frac{5}{7} - 1\frac{1}{3}$ p $1\frac{1}{2} - \frac{7}{3}$

   q $2\frac{1}{3} - \frac{17}{3}$ r $1\frac{4}{9} - \frac{13}{3}$ s $2\frac{1}{3} - \frac{12}{7}$ t $9\frac{1}{4} - \frac{17}{3}$

   TIP
   You can use the LCM of the denominators to find a common denominator, but any common denominator will work, it does not need to be the lowest one.

5. Calculate.

   a $8 \div \frac{1}{3}$ b $12 \div \frac{7}{8}$ c $\frac{7}{8} \div 12$

   d $\frac{2}{9} \div \frac{18}{30}$ e $\frac{8}{9} \div \frac{4}{5}$ f $1\frac{3}{7} \div 2\frac{2}{9}$

6. Simplify the following.

   a $4 + \frac{2}{3} \times \frac{1}{3}$ b $2\frac{1}{8} - \left(2\frac{1}{5} - \frac{7}{8}\right)$ c $\frac{3}{7} \times \left(\frac{2}{3} + 6 \div \frac{2}{3}\right) + 5 \times \frac{2}{7}$

   d $2\frac{7}{8} + \left(8\frac{1}{4} - 6\frac{3}{8}\right)$ e $\frac{5}{6} \times \frac{1}{4} + \frac{5}{8} \times \frac{1}{3}$ f $\left(5 \div \frac{3}{11} - \frac{5}{12}\right) \times \frac{1}{6}$

   g $\left(\frac{5}{8} \div \frac{15}{4}\right) - \left(\frac{5}{6} \times \frac{1}{5}\right)$ h $\left(2\frac{2}{3} \div 4 - \frac{3}{10}\right) \times \frac{3}{17}$ i $\left(7 \div \frac{2}{9} - \frac{1}{3}\right) \times \frac{2}{3}$

7 Jazmin has $900 in her account. She spends $\frac{7}{12}$ of this.
   a How much does she spend?
   b How much does she have left?

8 It takes a builder $\frac{3}{4}$ of an hour to lay 50 tiles.
   a How many tiles will the builder lay in $4\frac{1}{2}$ hours at the same rate?
   b If the builder lays tiles at the same rate for $6\frac{3}{4}$ hours a day, five days a week, how many tiles will they lay during the week?

9 $\frac{2}{5}$ of the people at an event ordered a sandwich. $\frac{1}{5}$ of the sandwiches ordered were vegetarian. What fraction of the people as the event ordered a vegetarian sandwich?

5.3 Percentages

KEY LEARNING STATEMENTS

- Per cent means per hundred. A percentage is a fraction with a denominator of 100.
- To find a percentage of a quantity, multiply the percentage by the quantity.
- To write one quantity as a percentage of another, express it as a fraction and then convert to a percentage by multiplying by 100.
- To increase or decrease an amount by a percentage, find the percentage amount and add or subtract it from the original amount.

KEY CONCEPTS

- Calculating a given percentage of a quantity.
- Expressing one quantity as a percentage of another.
- Working out percentage increase or decrease.

1 Express the following as percentages. Round your answers to one decimal place.
   a $\frac{1}{2}$ b $\frac{2}{3}$ c $\frac{1}{6}$ d $\frac{5}{8}$ e $\frac{93}{312}$
   f 0.3   g 0.04   h 0.47   i 1.12   j 2.07

2 Express the following percentages as common fractions in their simplest form.
   a 25%   b 80%   c 90%   d 12.5%
   e 50%   f 98%   g 60%   h 22%

3 Calculate.
   a 30% of 200 kg   b 40% of $60   c 25% of 600 litres
   d 22% of 250 ml   e 50% of $128   f 65% of $30
   g 215% of 120 km   h 0.5% of 40 grams   i 2.6% of $80
   j 9.5% of 5000 m³   k 2.5% of $80   l 120% of 3.5 kg

TIP

When you find a percentage of a quantity, your answer will have a unit and not a percentage sign because you are working out an amount.

4 Express the first quantity as a percentage of the second.

   a 400 metres of a 1000 metre race.

   b $4.80 out of a total of $240.

   c 27 kg removed from a 50 kg load.

   d 24 out of a total of 60 runs.

   e 25 out of 80 people.

5 Calculate the percentage increase or decrease.
   Round your answers to one decimal place.

	Original amount	New amount	Percentage increase or decrease
a	40	48	
b	4000	3600	
c	1.5	2.3	
d	12 000	12 400	
e	12 000	8600	
f	9.6	12.8	
g	90	2400	

6 Increase each amount by the given percentage.

   a $48 increased by 14%   b $700 increased by 35%

   c $30 increased by 7.6%   d $40 000 increased by 0.59%

   e $90 increased by 9.5%   f $80 increased by 24.6%

7 Decrease each amount by the given percentage.

   a $68 decreased by 14%   b $800 decreased by 35%

   c $90 decreased by 7.6%   d $20 000 decreased by 0.59%

   e $85 decreased by 9.5%   f $60 decreased by 24.6%

8 75 250 tickets were available for an international cricket match. 62% of the tickets were sold within a day. How many tickets are left?

9 Rajah owns 15% of a company. If the company issues 12 000 shares, how many shares should Rajah get?

10 A building, which cost $125 000 to build, increased in value by $3\frac{1}{2}$%.
   What is the building worth after this increase in value?

11 A player scored 18 out of the 82 points in a basketball match. What percentage of the points did that player score?

12 A company has a budget of $24 000 for printing brochures. The marketing department has already spent 34.6% of the budget. How much money is left in the budget?

13 A worker currently earns $6000 per month. If the worker receives an increase of 3.8%, what will their new monthly earnings be?

14 A company advertises that its cottage cheese is 99.5% fat free. If this is correct, how many grams of fat would there be in a 500 gram tub of the cottage cheese?

15 Li earns $25 per shift. The boss offered Li a choice between either $7 more per shift or a 20% increase. Which offer will give her the most money?

16 A cough contains about 3000 droplets and a sneeze can contain 40 000 droplets. A surgical mask will prevent most of these droplets from entering the air, but scientists estimate that 0.19% of the droplets will escape by passing through the mask even when it is properly and securely worn.

  a How many droplets will escape if you sneeze into a surgical mask?

  b If a person is sick, a single cough can contain 200 million individual virus particles. How many viral particles would escape from a surgical mask if the sick person coughed into it?

REFLECTION

Read the problem and look at the two students' answers.

Which student has the correct answer?

Why did the other student get the wrong answer?

After running 1800 metres of a 5 km race a runner stops to tie their shoelaces.

What percentage of the race has the athlete completed at this stage?

Student A
$$\frac{1800}{5000} \times 100 = \frac{9}{25} \times 100$$
$$= 36\%$$

Student B
$$\frac{1800}{5} \times 100 = 360 \times 100$$
$$= 360\%$$

5.4 Standard form

KEY LEARNING STATEMENTS

- A number in standard form is written as $A \times 10^n$, where $1 \leq A < 10$ and n is an integer.
- Standard form is also called scientific notation.
- To write a number in standard form:
  - First place a decimal point after the first significant digit.
  - Then count the number of places this first significant digit must move to get back to the original number. This gives the power of 10.
  - If the original number was smaller than the new number then the power of 10 is negative. If the original number was larger than the new number then the power of 10 is positive.
- To write a number in standard form as an ordinary number, multiply the decimal by 10 to the given power.
- When you multiply numbers in standard form you add the indices, when you divide numbers in standard form you subtract the indices. Remember to give the answer in standard form.

KEY CONCEPTS

- Converting numbers into and out of standard form ($A \times 10^n$).
- Calculating with values in standard form.

1 Write the following numbers in standard form.

a 45 000
b 800 000
c 80
d 2 345 000
e 4 190 000
f 32 000 000 000
g 0.0065
h 0.009
i 0.000 45
j 0.000 000 8
k 0.006 75
l 0.000 000 000 45

TIP
Make sure you know how your calculator deals with standard form.

2 Write the following as ordinary numbers.

a 2.5×10^3
b 3.9×10^4
c 4.265×10^5
d 1.045×10^{-5}
e 9.15×10^{-6}
f 1×10^{-9}
g 2.8×10^{-5}
h 9.4×10^7
i 2.45×10^{-3}

TIP
If the number part of your standard form answer is a whole number, there is no need to add a decimal point.

3 Calculate, giving your answers in standard form correct to three significant figures.

a 4216^6
b $(0.000\,09)^4$
c $0.0002 \div 2500^3$
d $65\,000\,000 \div 0.000\,004\,5$
e $(0.0029)^3 \times (0.003\,65)^5$
f $(48 \times 987)^4$
g $\dfrac{4525 \times 8760}{0.00002}$
h $\dfrac{9500}{0.0005^4}$
i $\sqrt{5.25 \times 10^8}$

4   Simplify each of the following. Give your answer in standard form.

   a $(3 \times 10^{12}) \times (4 \times 10^{18})$
   b $(1.5 \times 10^6) \times (3 \times 10^5)$
   c $(1.5 \times 10^{12})^3$
   d $(1.2 \times 10^{-5}) \times (1.1 \times 10^{-6})$
   e $(0.4 \times 10^{15}) \times (0.5 \times 10^{12})$
   f $(8 \times 10^{17}) \div (3 \times 10^{12})$
   g $(1.44 \times 10^8) \div (1.2 \times 10^6)$
   h $(8 \times 10^{-15}) \div (4 \times 10^{-12})$
   i $\sqrt[3]{27} \times 10^{-8}$

5 The Sun has a mass of approximately 1.998×10^{27} tonnes. The planet Mercury has a mass of approximately 3.302×10^{20} tonnes.

   a Which has the greater mass?
   b How many times heavier is the greater mass compared with the smaller mass?

6 Light travels at a speed of 3×10^8 metres per second. The Earth is an average distance of 1.5×10^{11} metres from the Sun and Pluto is an average 5.9×10^{12} metres from the Sun.

   a Work out how long it takes light from the Sun to reach Earth (in seconds). Give your answer as both an ordinary number and a number in standard form.
   b How much longer does it take for the light to reach Pluto? Give your answer as both an ordinary number and a number in standard form.

REVIEW EXERCISE

1 Simplify.

   a $\dfrac{160}{200}$
   b $\dfrac{48}{72}$
   c $\dfrac{36}{54}$

2 Calculate.

   a $\dfrac{4}{9} \times \dfrac{3}{8}$
   b $84 \times \dfrac{3}{4}$
   c $\dfrac{5}{9} \div \dfrac{1}{3}$
   d $\dfrac{4}{15} + \dfrac{9}{15}$
   e $\dfrac{9}{11} - \dfrac{3}{4}$
   f $\dfrac{5}{24} + \dfrac{7}{16}$
   g $2\dfrac{1}{3} + 9\dfrac{1}{2}$
   h $\left(4\dfrac{3}{4}\right)^2$
   i $9\dfrac{1}{5} - 1\dfrac{7}{9}$

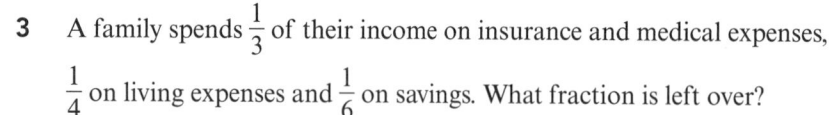

3 A family spends $\dfrac{1}{3}$ of their income on insurance and medical expenses, $\dfrac{1}{4}$ on living expenses and $\dfrac{1}{6}$ on savings. What fraction is left over?

4 Express the first quantity as a percentage of the second.

   a $400, $5000
   b 4.8, 96
   c 19, 30

5 A traffic officer pulls over 12 drivers at random from 450 passing cars. What percentage of drivers is this?

> **CONTINUED**
>
> **6** Find:
>
>   **a** 30% of 82 kg   **b** 2.5% of 20 litres   **c** 17.5% of $400.
>
> **7** Express the following as percentages.
>
>   **a** $\frac{1}{8}$ **b** $\frac{1}{3}$ **c** 425 out of 1250
>
> **8** Increase $90 by 15%.
>
> **9** Decrease $42.50 by 12%.
>
> **10** A baby had a mass of 3.25 kg at birth. After 12 weeks, the baby's mass had increased to 5.45 kg. Express this as a percentage increase, correct to one decimal place.
>
> **11** In general, a customs officer checks the luggage of 18 randomly selected passengers out of 720 on a flight.
>
>   **a** What percentage of passengers do not have their luggage checked?
>
>   **b** Of those checked, 75% of the passengers are allowed to proceed, the others have to go for further security screening. What percentage of the passengers on the flight end up going for further screening?
>
> **12** An aeroplane was flying at an altitude of 10 500 metres when it experienced serious turbulence and dropped 28% of its altitude. How many metres did it drop?
>
> **13** Pluto is 5.9×10^{12} metres from the Sun.
>
>   **a** Express this in kilometres, giving your answer in standard form.
>
>   **b** In a certain position, the Earth is 1.47×10^{8} km from the Sun. If Pluto, the Earth and the Sun are in a straight line in this position (and both planets are the same side of the sun), calculate the approximate distance, in kilometres, between the Earth and Pluto. Give your answer in standard form.

> Chapter 6: Equations, factors and formulae

6.1 Solving equations

KEY LEARNING STATEMENTS

- To solve an equation, you find the value of the unknown letter (variable) that makes the equation true.
- If you add or subtract the same number (or term) to both sides of the equation, you produce an equivalent equation and the solution remains unchanged.
- If you multiply or divide each term on both sides of the equation by the same number (or term), you produce an equivalent equation and the solution remains unchanged.

KEY CONCEPT

Solving equations with one unknown.

In this exercise, leave answers as fractions rather than decimals where necessary.

1 Solve these equations.

   a $x + 5 = 21$ **b** $x - 10 = 14$ **c** $4x = 32$ **d** $\dfrac{x}{6} = 9$

   e $9x = 63$ **f** $x - 2 = -4$ **g** $x + 7 = -9$ **h** $\dfrac{x}{5} = -12$

   i $x - 4 = -13$ **j** $-4x = 60$ **k** $-2x = -26$ **l** $-3x = -45$

2 Solve these equations for x. Show the steps in your working.

   a $2x + 3 = 19$ **b** $3x - 9 = 36$ **c** $2x + 9 = 4$

   d $4 - 2x = 24$ **e** $-4x + 5 = 21$ **f** $-2x - 9 = 15$

3 Solve these equations for x. Show the steps in your working.

   a $2x + 7 = 3x + 4$ **b** $4x + 6 = x + 18$ **c** $5x - 2 = 3x + 7$

   d $9x - 5 = 7x + 3$ **e** $11x - 4 = x + 32$ **f** $2x - 1 = 14 - x$

   g $20 - 4x = 5x + 2$ **h** $3 + 4x = 2x - 7$ **i** $4x + 5 = 7x - 7$

   j $2x - 6 = 4x - 3$ **k** $3x + 2 = 5x - 9$ **l** $x + 9 = 5x - 3$

TIP

The variable can appear on both sides of the equation. You can add or subtract variables to both sides just like numbers.

6 Equations, factors and formulae

4 Solve these equations for x.

a $3(x - 2) = 24$
b $5(x + 4) = 10$
c $3(3x + 10) = 6$
d $3(2x - 1) = 5$
e $-3(x - 6) = -6$
f $4(3 - 5x) = 7$
g $4(x + 3) = x$
h $6(x + 3) = 4x$
i $3x + 2 = 2(x - 4)$
j $x - 3 = 2(x + 5)$
k $4(x + 7) - 3(x - 5) = 9$
l $2(x - 1) - 7(3x - 2) = 7(x - 4)$

TIP

When an equation has brackets it is often best to expand them first.

5 Solve these equations for x.

a $\dfrac{x}{2} - 3 = 6$
b $\dfrac{x}{3} + 2 = 11$
c $\dfrac{4x}{6} = 16$
d $\dfrac{28 - x}{6} = 12$
e $\dfrac{x - 2}{3} = 5$
f $\dfrac{x + 3}{2} = 16$
g $\dfrac{2x - 5}{3} = 9$
h $\dfrac{2x - 1}{5} = 9$
i $\dfrac{5x + 2}{3} = -1$
j $\dfrac{5 - 2x}{4} = 1$
k $\dfrac{2x - 1}{5} = x$
l $\dfrac{2x - 3}{5} = x - 6$
m $\dfrac{10x + 2}{3} = 6 - x$
n $\dfrac{x}{2} - \dfrac{x}{5} = 3$
o $\dfrac{2x}{3} - \dfrac{x}{2} = 7$
p $-2\dfrac{(x + 4)}{2} = x + 7$

TIP

To remove the denominators of fractions in an equation, multiply each term on both sides by the common denominator.

6.2 Factorising algebraic expressions

KEY LEARNING STATEMENTS

- The first step in factorising is to identify and 'take out' **all** common factors.
- Common factors can be numbers, variables, brackets or a combination of these.
- Factorising is the opposite of expanding – when you factorise you put brackets back into the expression.

KEY CONCEPT

Extracting common factors.

1 Find the highest common factor of each pair.

a $3x$ and 21
b 40 and $8x$
c $15a$ and $5b$
d $2a$ and ab
e $3xy$ and $12yz$
f $5a^2b$ and $20ab^2$
g $8xy$ and $28xyz$
h $9pq$ and p^2q^2
i $14abc$ and $7a^2b$
j x^2y^3z and $2xy^2z^2$
k $2a^2b^4$ and ab^3
l $3x^3y^2$ and $15xy$

TIP

Remember, x^2 means $x \times x$, so x is a factor of x^2.

2 Factorise as fully as possible.

   a $12x + 48$ b $2 + 8y$ c $4a - 16$ d $2x - 12$
   e $4x - 20$ f $16a - 8$ g $3x - xy$ h $ab + 5a$
   i $3x - 15y$ j $8a + 24$ k $12x - 18$ l $24xyz - 8xz$
   m $9ab - 12bc$ n $6xy - 4yz$ o $14x - 26xy$ p $-14x^2 - 7x^5$

TIP

Find the HCF of the numbers first. Then find the HCF of the variables, if there is one, in alphabetical order.

3 Factorise the following.

   a $x^2 + 8x$ b $12a - a^2$ c $9x^2 + 4x$
   d $22x - 16x^2$ e $6ab^2 + 8b$ f $18xy - 36x^2y$
   g $6x - 9x^2$ h $14x^2y^2 - 6xy^2$ i $9abc^3 - 3a^2b^2c^2$
   j $4x^2 - 7xy$ k $3ab^2 - 4b^2c$ l $14a^2b - 21ab^2$

TIP

Remember, if one of the terms is exactly the same as the common factor, you must put a 1 where the term would appear in the bracket.

4 Remove a common factor to factorise each of the following expressions.

   a $x(3 + y) + 4(3 + y)$ b $x(y - 3) + 5(y - 3)$
   c $3(a + 2b) - 2a(a + 2b)$ d $4a(2a - b) - 3(2a - b)$
   e $x(2 - y) + (2 - y)$ f $x(x - 3) + 4(x - 3)$
   g $9(2 - y) - x(y - 2)$ h $4a(2b - c) - (c - 2b)$
   i $3x(x - 6) - 5(x - 6)$ j $x(x - y) - (2x - 2y)$
   k $3x(2x + 3) + y(3 + 2x)$ l $4(x - y) - x(3x - 3y)$

6.3 Rearranging formulae

KEY LEARNING STATEMENTS

- A formula is a general rule, usually involving several variables, for example, the area of a rectangle, $A = lw$.
- A variable is called the subject of the formula when it is on its own on one side of the equals sign.
- You can rearrange a formula to make any variable the subject. You use the same rules that you used to solve equations.

KEY CONCEPT

Rearranging simple formulae.

1 Make m the subject if $D = km$

2 Make c the subject if $y = mx + c$

3 Given that $P = ab - c$, make b the subject of the formula.

4 Given that $a = bx + c$, make b the subject of the formula.

6 Equations, factors and formulae

5 Make a the subject of each formula.

   a $a + b = c$ **b** $a - 3b = 2c$ **c** $ab - c = d$ **d** $ab + c = d$

   e $bc - a = d$ **f** $bc - a = -d$ **g** $\dfrac{2a + b}{c} = d$ **h** $\dfrac{c + ba}{d} = e$

   i $abc - d = e$ **j** $cab + d = ef$ **k** $\dfrac{ab}{c} + de = f$ **l** $c + \dfrac{ab}{d} = e$

   m $c(a - b) = d$ **n** $d(a + 2b) = c$

> **TIP**
> Pay attention to the signs when you rearrange a formula.

6 The perimeter of a rectangle can be given as $P = 2(l + w)$, where P is the perimeter, l is the length and w is the width.

   a Make w the subject of the formula.

   b Find w if the rectangle has a length of 45 cm and a perimeter of 161 cm.

7 The circumference of a circle can be found using the formula $C = 2\pi r$, where r is the radius of the circle.

   a Make r the subject of the formula.

   b Find the radius of a circle of circumference 56.52 cm. Use $\pi = 3.14$.

   c Find the diameter of a circle of circumference 144.44 cm. Use $\pi = 3.14$.

> **TIP**
> If you are given a value for π, you must use the given value to avoid calculator and rounding errors.

> **TIP**
> In questions such as Q8, it may be helpful to draw a diagram to show what the parts of the formula represent.

8 The area of a trapezium can be found using the formula $A = \dfrac{h(a + b)}{2}$, where h is the distance between the parallel sides and a and b are the lengths of the parallel sides. By rearranging the formula and substitution, find the length of b in a trapezium of area 9.45 cm² with $a = 2.5$ cm and $h = 3$ cm.

REVIEW EXERCISE

1 Solve for x.

   a $4x - 9 = -21$ **b** $5x + 4 = -26$

   c $\dfrac{2x - 4}{7} = 2$ **d** $5 = \dfrac{1 - 4x}{5}$

   e $4x - 6 = 12 - 5x$ **f** $4x - 8 = 3(2x + 6)$

   g $\dfrac{3x - 7}{4} = \dfrac{1 - 4x}{8}$ **h** $\dfrac{3(2x - 5)}{5} = \dfrac{x + 1}{2}$

2 Make x the subject of each formula.

   a $m = nxp - r$ **b** $m = \dfrac{nx + p}{q}$

CONTINUED

3 Factorise fully.

  a $4x - 8$ **b** $12x - 3y$

  c $-2x - 4$ **d** $3xy - 24x$

  e $14x^2y^2 + 7xy$ **f** $2(x - y) + x(x - y)$

  g $x(4 + 3x) - 3(3x + 4)$ **h** $4x^2(x + y) - 8x(x + y)$

4 Given that, for a rectangle, area = length × width, write an expression for the area of each rectangle. Expand each expression fully.

  a rectangle with length $x - 7$ and width 4

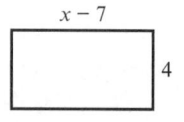

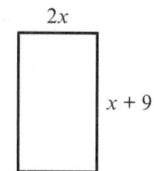

  b rectangle with width $2x$ and length $x + 9$

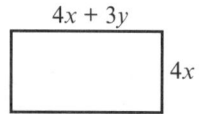

  c rectangle with length $4x + 3y$ and width $4x$

  d rectangle with length $19x$ and width $x + 2y$

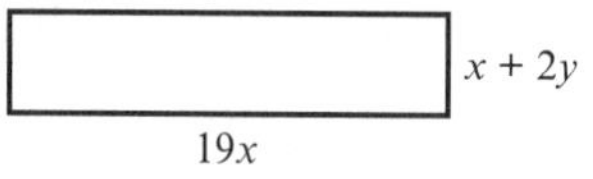

SELF ASSESSMENT

How well did you do in this Review exercise?

1 Use this checklist to rate your own work.
List the letters A to F in your book and write one of these symbols next to each one:

★★★ – I did really well

★★ – I tried and mostly succeeded

★ – This needs more work

Checklist

  A Solve simple equations.

  B Solve equations with fractions in them.

  C Identify common factors in algebraic expressions.

  D Factorise expressions where the common factor is a bracket.

  E Rearrange formulae to make any letter the subject of the formula.

  F Substitute values and use formulae to solve problems.

2 Decide what you will do about any areas that need more work.

Chapter 7: Perimeter, area and volume

7.1 Perimeter and area in two dimensions

KEY LEARNING STATEMENTS

- Perimeter is the total distance around the outside of a shape. You can find the perimeter of any shape by adding up the lengths of the sides.
- The perimeter of a circle is called the circumference. Use the formula $C = \pi d$ or $C = 2\pi r$ to find the circumference of a circle.
- Area is the total space contained within a shape. Use these formulae to calculate the area of different shapes:
    - triangle: $A = \dfrac{bh}{2}$
    - square: $A = s^2$
    - rectangle: $A = bh$
    - parallelogram: $A = bh$
    - rhombus: $A = bh$
    - kite: $A = \dfrac{1}{2}$ (product of diagonals)
    - trapezium: $A = \dfrac{(\text{sum of parallel sides}) \times h}{2}$
    - circle: $A = \pi r^2$
- You can work out the area of compound shapes in a few steps.
    - Divide compound shapes into known shapes.
    - Work out the area of each part and then add the areas together to find the total area.
- You can find the length of an arc and the perimeter or area of sectors of a circle using the angle at the centre.
    - Arc length is $\dfrac{\text{angle at centre}}{360} \times 2\pi r$
    - Area of a sector is $\dfrac{\text{angle at centre}}{360} \times \pi r^2$

KEY CONCEPTS

- Calculating perimeter and area of polygons and compound shapes.
- Calculating circumference and area of a circle.
- Solving problems involving sectors and arc lengths of circles.

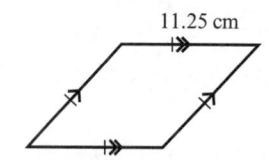

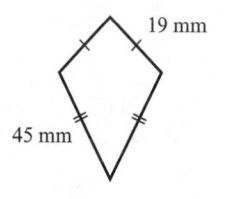

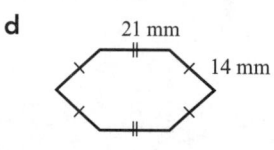

1 Find the perimeter of each shape.

a 32 mm, 28 mm (parallelogram)

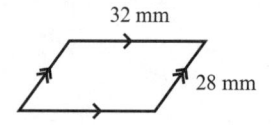

b 11.25 cm (rhombus)

c 19 mm, 45 mm (kite)

d 21 mm, 14 mm (hexagon)

e 1.5 cm, 5.3 cm, 6.8 cm, 3.4 cm, 4.9 cm (compound shape)

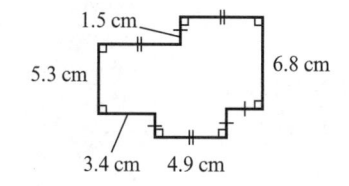

f 92 mm, 7.2 cm, 69 mm (triangle)

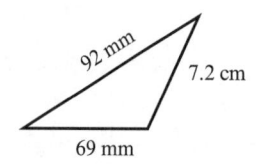

2 Find the perimeter of each of these shapes. Give your answers correct to two decimal places.

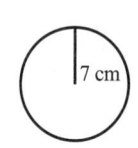

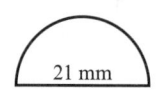

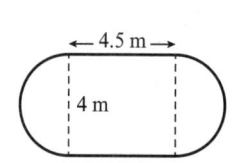

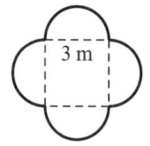

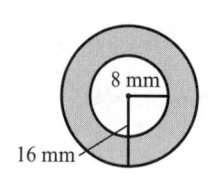

a circle, diameter 5 m

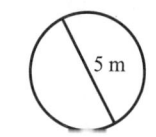

b circle, radius 7 cm

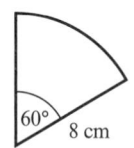

c semicircle, diameter 21 mm

d 4.5 m, 4 m (stadium/rectangle with semicircular ends)

e 3 m (four semicircles on a square)

f 8 mm, 16 mm (annulus)

g 60°, 8 cm (sector)

h 6 cm, 5 cm, 2 cm, 6 cm (shape)

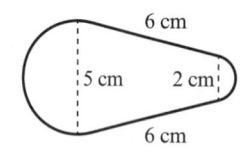

TIP
If you are not given the value of π, use the π key on your calculator, unless you are asked to give your answer as a multiple of π.

TIP
Drawing a rough sketch can help you work out what is required and how to find it.

3 A square field has a perimeter of 360 metres. What is the length of one of its sides?

4 Find the cost of fencing a rectangular plot 45 metres long and 37 metres wide if the cost of fencing is $45.50 per metre.

5 An isosceles triangle has a perimeter of 28 cm. Calculate the length of each of the equal sides if the third side is 100 mm long.

6 How much string do you need to form a circular loop with a diameter of 28 cm?

7 The rim of a bicycle wheel has a radius of 31.5 cm.

  a What is the circumference of the rim? Give your answer as an exact multiple of π.

  b The tyre that goes onto the rim is 3.5 cm thick. Calculate the circumference of the wheel when the tyre is fitted to it. Give your answer as an exact multiple of π.

8 Find the area of each of these shapes.

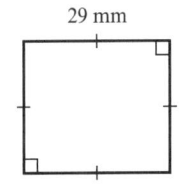

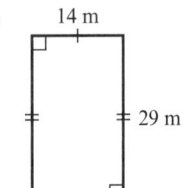

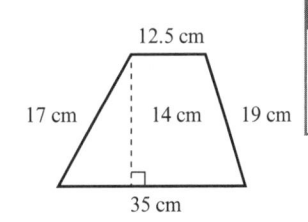

a square, side 29 mm

b rectangle, 14 m × 29 m

c trapezium, parallel sides 12.5 cm and 35 cm, slant sides 17 cm and 19 cm, height 14 cm

d parallelogram, base 1.7 m, height 90 cm

e kite, diagonals 12 cm + (unmarked) horizontal and 19 cm vertical

f parallelogram, base 21 cm, height 19 cm

g right-angled triangle, legs 15 cm and 20 cm, hypotenuse 25 cm

h isosceles triangle, base 17 cm, equal sides 11 cm, height 7 cm

i triangle with 11 cm, 21 cm, 12 cm, 5 cm, 13 cm

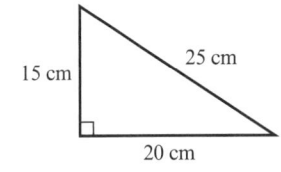

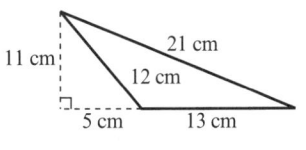

j triangle, sides 72 mm, 112 mm, 67 mm, with 41 mm perpendicular

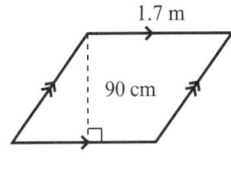

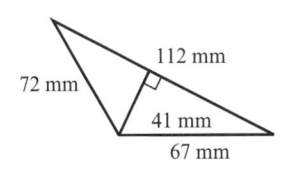

k square, side 2.4 m

l rectangle, 147.6 cm × 49.2 cm

m kite, 8 cm, 10 cm, 15 cm, 12 cm

> **TIP**
> Make sure all measurements are in the same units before you do any calculations.

> **TIP**
> Remember, give your answer in square units.

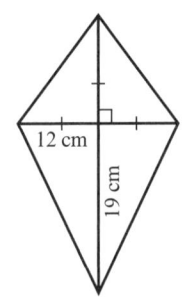

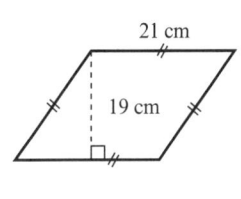

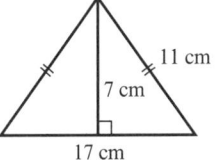

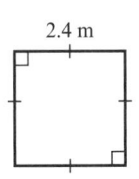

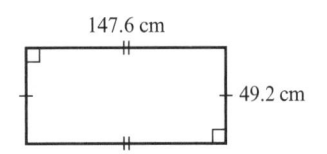

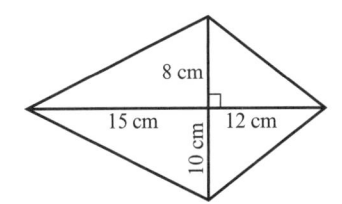

9 Find the area of each shape. Give your answers correct to two decimal places.

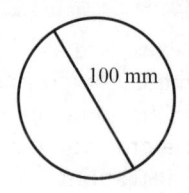

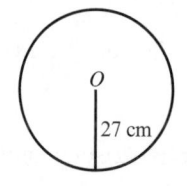

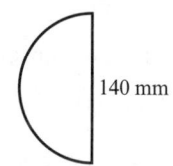

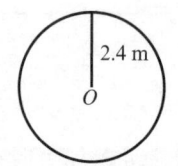

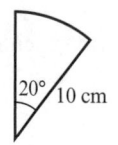

a 100 mm

b 27 cm

c 140 mm

d 2.4 m

e 20° 10 cm

10 Find the area of each of these figures. Show your working clearly in each case and give your answers to two decimal places where necessary. All dimensions are given in centimetres.

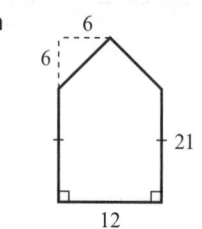

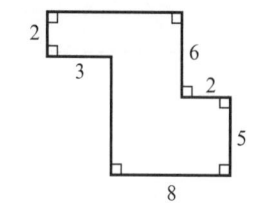

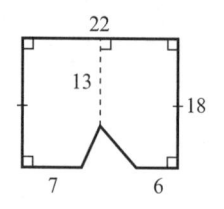

> **TIP**
>
> Work out any missing dimensions on the figure using the given dimensions and the properties of shapes.

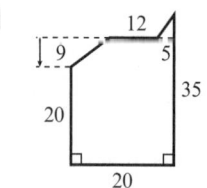

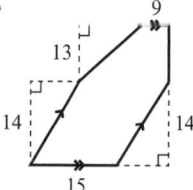

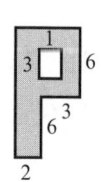

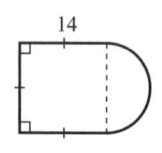

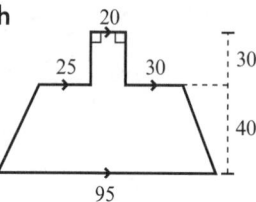

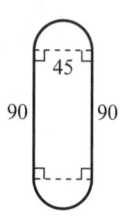

a 6, 6, 21, 12

b 2, 3, 6, 2, 5, 8

c 22, 13, 18, 7, 6

d 12, 9, 5, 35, 20, 20

e 9, 13, 14, 14, 15

f 1, 3, 6, 3, 6, 2

g 14

h 20, 25, 30, 30, 40, 95

i 45, 90, 90

11 Find the area of the following figures giving your answers correct to two decimal places. (Give your answer to part (f) in terms of π.)

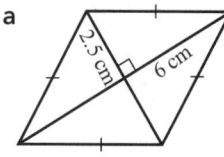

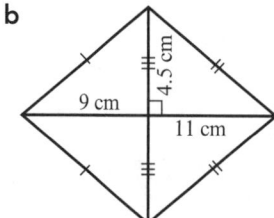

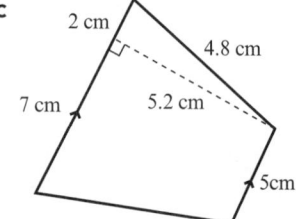

a 2.5 cm, 6 cm

b 9 cm, 4.5 cm, 11 cm

c 2 cm, 4.8 cm, 7 cm, 5.2 cm, 5 cm

> **TIP**
>
> Divide irregular shapes into known shapes and combine the areas to get the total area.

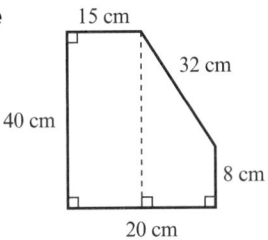

d (quadrilateral with sides 6.5 cm, 4 cm, 3 cm, 5.5 cm, 6.3 cm)

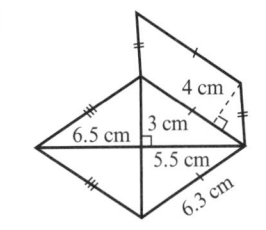

e (shape with 15 cm top, 32 cm, 40 cm left, 8 cm, 20 cm bottom)

f (circle with 50 mm, 40 mm, 30 mm)

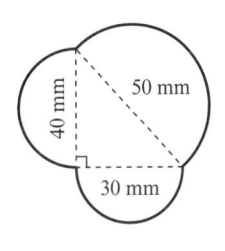

12 A 1.5 metre × 2.4 metre rectangular rug is placed on the floor in a 3.5 metre × 4.2 metre rectangular room. How much of the floor is not covered by the rug?

TIP

Draw a diagram and add all the information you are given.

13 The area of a rhombus of side 8 cm is 5600 mm². Determine the height of the rhombus.

14 Calculate the length of the arc AB in each of these circles. Give your answers correct to two decimal places.

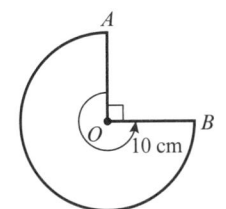

a (circle, radius 21 mm, angle 120°)

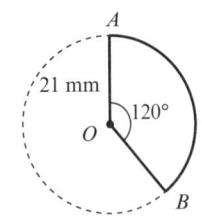

b (circle, radius 10 cm, reflex angle at O)

c (circle, radius 12 mm, angle 40°)

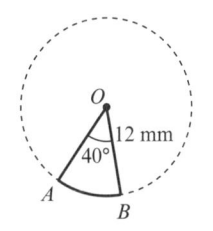

15 The diagram shows a cross-section of the Earth. Two cities, X and Y, lie on the same longitude. Given that the radius of the Earth is 6371 km, calculate the distance, XY, between the two cities. Give your answer correct to two decimal places.

(circle with radius 6371 km, angle 60°)

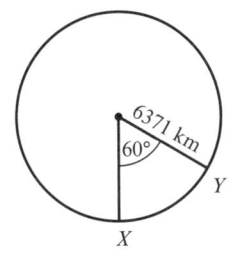

16 Calculate the shaded area of each circle. Give your answers as a multiple of π.

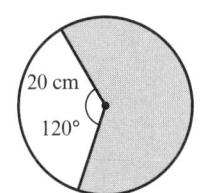

a (circle radius 12 cm, angle 60°)

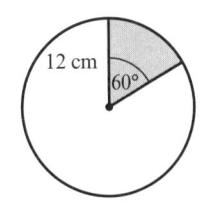

b (circle radius 20 cm, angle 120°)

c (circle radius 18 mm)

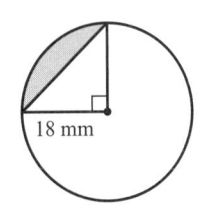

17 A large circular pizza has a diameter of 25 cm. The pizza restaurant cuts its pizzas into eight equal slices. Calculate the area of each slice in cm² correct to three significant figures.

7.2 Three-dimensional objects

> **KEY LEARNING STATEMENTS**
>
> - Any solid object is three-dimensional. The three dimensions of a solid are length, width and height.
> - The net of a solid is a two-dimensional diagram. It shows the shape of all faces of the solid and how they are attached to each other. If you fold up a net, you get a model of the solid.

KEY CONCEPT

Solids and nets of solids.

1 Which solids can be made from the following sets of faces?

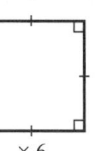

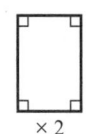

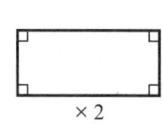

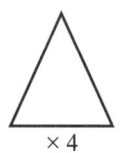

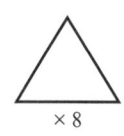

   a □ × 6

   b □ × 2   ▭ × 2   ▭ × 2

   c △ × 4   □ × 1

   d △ × 8

2 Describe the solid you can produce using each of the following nets.

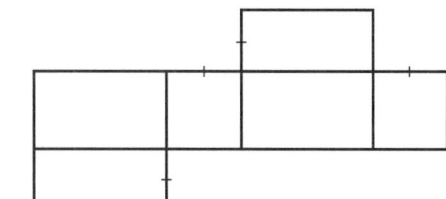

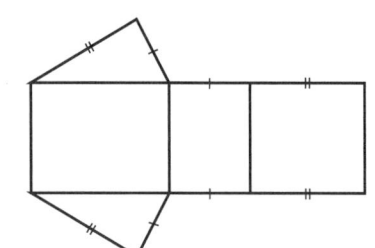

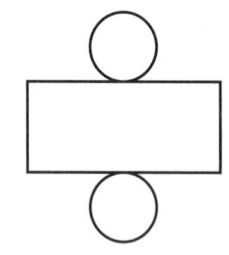

   a

   b

   c

3 Sketch a possible net for each of the following solids.

a

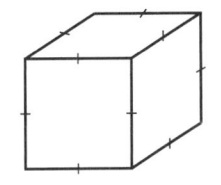

b

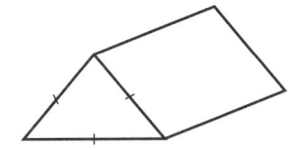

c

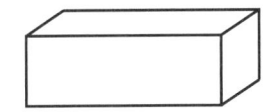

d

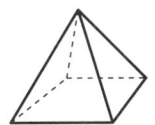

7.3 Surface areas and volumes of solids

KEY LEARNING STATEMENTS

- The surface area of a three-dimensional object is the total area of all its faces.
- The volume of a three-dimensional object is the amount of space it occupies.
- You can find the volume of a cube or cuboid using the formula $V = l \times w \times h$, where l is the length, w is the width and h is the height of the object.
- A prism is a three-dimensional object with a uniform cross-section, (the end faces of the solid are identical and parallel). If you slice through the prism anywhere along its length (and parallel to the end faces), you will get a cross-section the same shape and size as the end faces. Cubes and cuboids are prisms.
- You can find the volume of any prism or cylinder by multiplying the area of its cross-section by the distance between the parallel faces. This is expressed in the formula $V = al$, where a is the area of the base and l is the length of the prism. You need to use the appropriate area formula for the shape of the cross-section.
- Find the volume of a cone using the formula, $V = \frac{1}{3}\pi r^2 h$, where h is the perpendicular height and r is the radius of the circular base. To find the curved surface area use the formula, surface area = $\pi r l$, where l is the slant height of the cone.
- Find the volume of a pyramid using the formula, $V = \frac{\text{area of base} \times h}{3}$, where h is the perpendicular height.
- Find the volume of a sphere using the formula, $V = \frac{4}{3}\pi r^3$. To find the surface area use the formula, surface area = $4\pi r^2$.

KEY CONCEPTS

- Calculating surface area and volume of cuboids, prisms, cylinders, spheres, pyramid and cones.
- Calculating volume and surface area of compound shapes.

CAMBRIDGE IGCSE™ MATHEMATICS: CORE PRACTICE BOOK

1 Calculate the surface area of each shape. Give your answers to two decimal places where necessary.

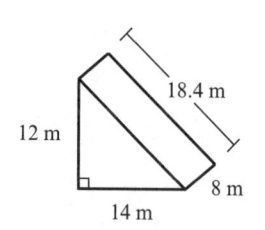

a 1.2 mm × 0.4 mm × 0.5 mm cuboid

b Triangular prism with sides 12 m, 14 m, 18.4 m and length 8 m

c Cube with side 1.5 cm

d Cylinder with radius 4 mm and height 12 mm

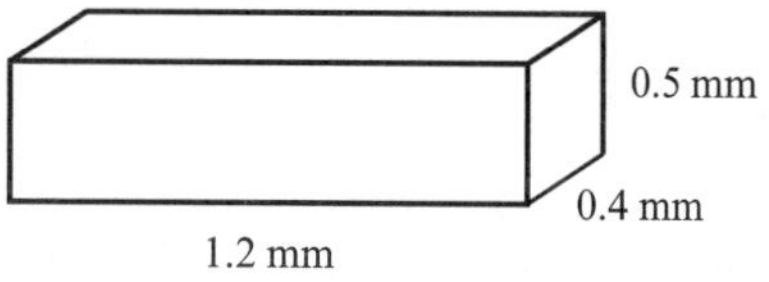

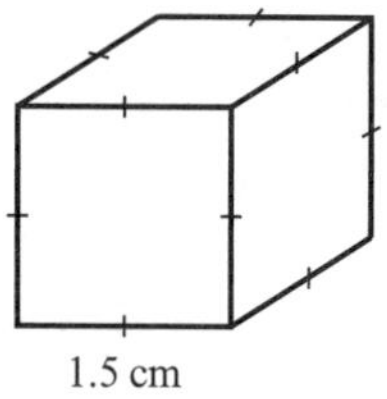

TIP

Drawing the nets of the shapes may help you work out the area of each face.

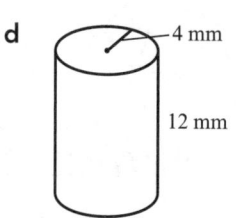

2 A wooden cube has six identical square faces, each of area 64 cm².

  a What is the surface area of the cube?

  b What is the height of the cube?

3 Mrs Nini is ordering wooden blocks to use in her maths classroom. The blocks are cuboids with dimensions 10 cm × 8 cm × 5 cm.

  a Calculate the surface area of one block.

  b Mrs Nini needs 450 blocks. What is the total surface area of all the blocks?

  c She decides to varnish the blocks. A tin of varnish covers an area of 4 m². How many tins will she need to varnish all the blocks?

TIP

Remember,
1 m² = 10 000 cm²

4 Calculate the volume of each prism. Give your answers to two decimal places where necessary.

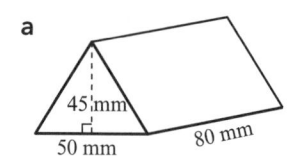

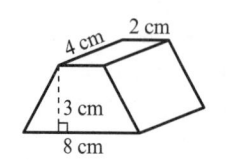

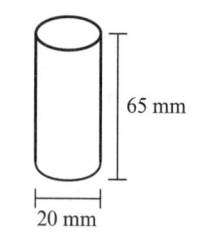

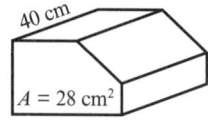

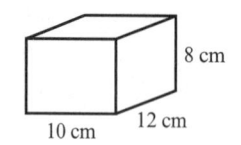

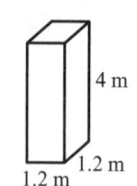

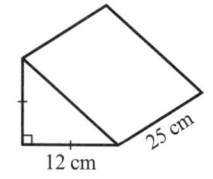

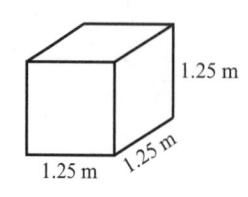

a Triangular prism: base 50 mm, height 45 mm, length 80 mm

b Trapezoidal prism: parallel sides 4 cm and 2 cm, height 3 cm, length 8 cm

c Cylinder: diameter 20 mm, height 65 mm

d Prism with cross-sectional area $A = 28$ cm², length 40 cm

e Cuboid 10 cm × 12 cm × 8 cm

f Cuboid 1.2 m × 1.2 m × 4 m

g Triangular prism: base 12 cm, hypotenuse 25 cm

h Cube with side 1.25 m

TIP

The length of the prism is the distance between the two parallel faces. When a prism is turned onto its face, the length may look like a height. Work out the area of the cross-section (end face) before you apply the volume formula.

60

7 Perimeter, area and volume

5 Find the volume of the following solids. Give your answers correct to two decimal places.

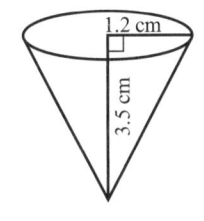

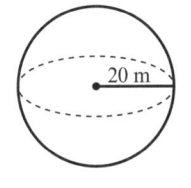

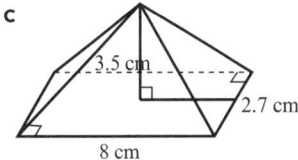

a. cone with diameter 1.2 cm and height 3.5 cm

b. sphere with diameter 20 m

c. pyramid with base 8 cm by 2.7 cm and height 3.5 cm

6 This shape was built using 2 cm cubes. What is its volume?

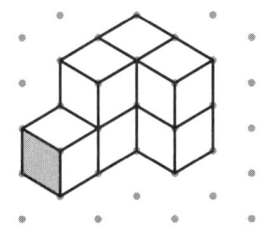

7 Find the volume and total surface area of the following compound solids. Give your answers correct to two decimal places.

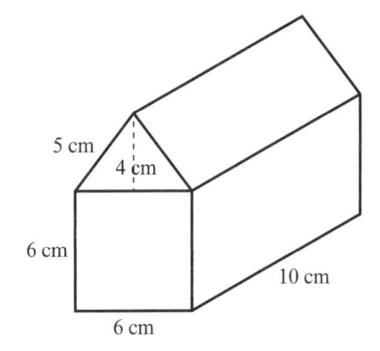

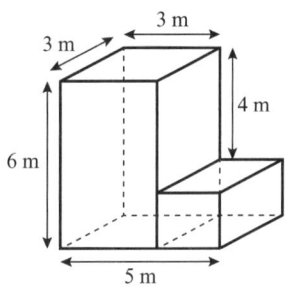

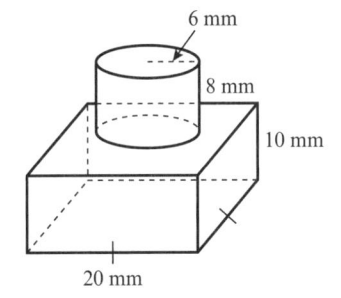

a. prism (house shape): 6 cm, 6 cm, 10 cm, with roof triangle height 4 cm and slant 5 cm

b. stepped solid: 6 m, 5 m, 3 m, 3 m, 4 m

c. cylinder (radius 6 mm, height 8 mm) on top of cuboid (20 mm × 20 mm × 10 mm)

8 A pocket dictionary is 14 cm long, 9.5 cm wide and 2.5 cm thick. Calculate the volume of space it takes up.

9 a Find the volume of a lecture room that is 8 metres long, 8 metres wide and 3.5 metres high.

b Safety regulations state that during an hour long lecture each person in the room must have 5 m³ of air. Calculate the maximum number of people who can attend an hour long lecture.

10 A cylindrical tank is 30 metres high with an inner radius of 150 cm. Calculate how much water the tank will hold when full. Give your answer in terms of π.

11 A machine shop has four different rectangular prisms of volume 64 000 mm^3. Copy and fill in the possible dimensions for each prism to complete the table.

Volume (mm^3)	64 000	64 000	64 000	64 000
Length (mm)	80	50		
Width (mm)	40		80	
Height (mm)				16

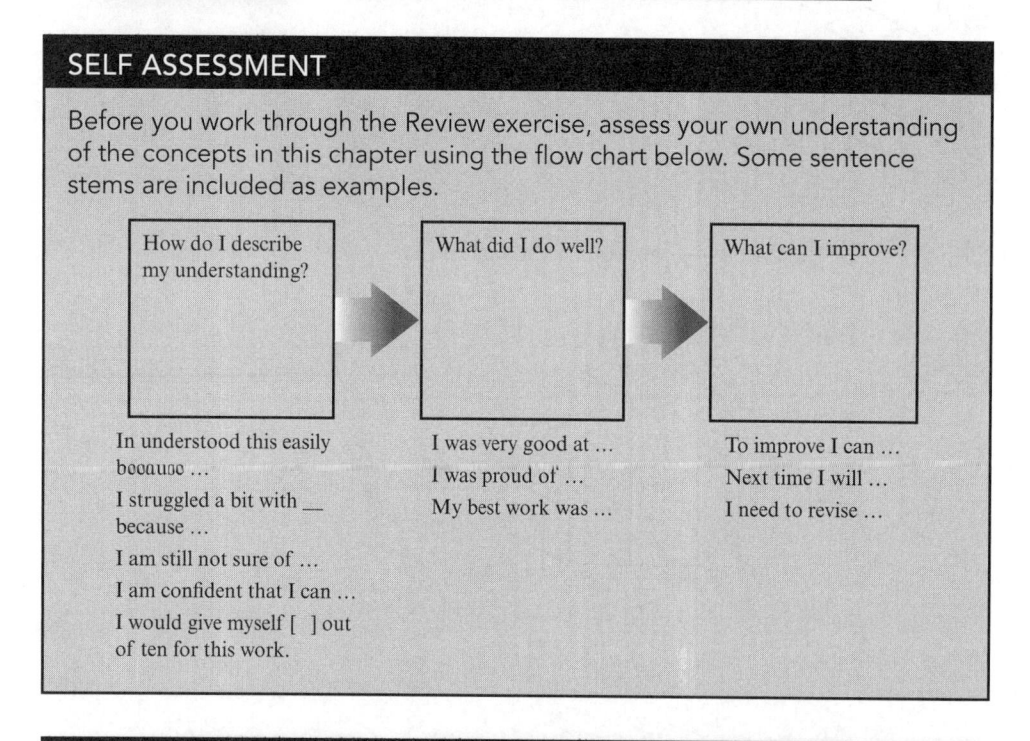

SELF ASSESSMENT

Before you work through the Review exercise, assess your own understanding of the concepts in this chapter using the flow chart below. Some sentence stems are included as examples.

How do I describe my understanding?

What did I do well?

What can I improve?

In understood this easily because ...
I struggled a bit with ___ because ...
I am still not sure of ...
I am confident that I can ...
I would give myself [ ] out of ten for this work.

I was very good at ...
I was proud of ...
My best work was ...

To improve I can ...
Next time I will ...
I need to revise ...

REVIEW EXERCISE

1 A circular plate on the top of an oven has a diameter of 21 cm. There is a metal strip around the outside of the plate.

   a Calculate the surface area of the top of the plate.

   b Calculate the length of the metal strip.

2 What is the radius of a circle with an area of 65 cm^2?

CONTINUED

3 Calculate the shaded area in each figure. Give your answers to two decimal places where necessary.

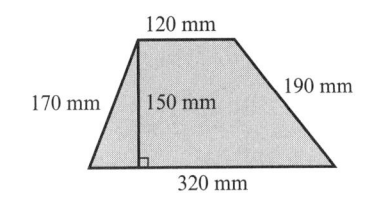

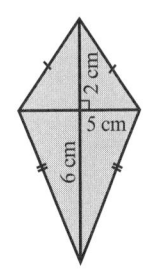

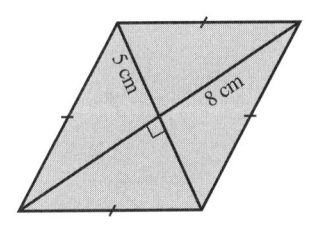

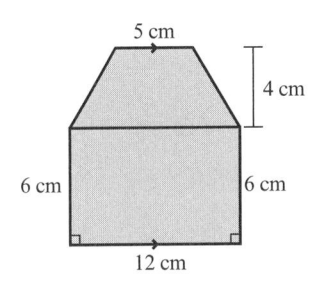

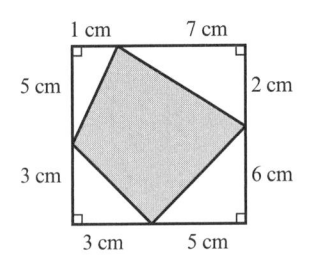

a parallelogram with base side 50 mm and perpendicular height 40 mm

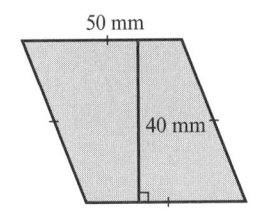

b trapezium with parallel sides 120 mm (top) and 320 mm (bottom), slant sides 170 mm and 190 mm, perpendicular height 150 mm

c kite with diagonals: half-diagonals 2 cm and 5 cm (vertical parts) and 6 cm with tick marks

d rhombus with diagonals 5 cm and 8 cm (half-diagonals shown)

e pentagon: top slant sides meeting at apex above a 5 cm top, height of triangle 4 cm, rectangle below 12 cm wide and 6 cm tall

f square of side 8 cm (1 cm + 7 cm across top; 5 cm + 3 cm down left; 2 cm + 6 cm down right; 3 cm + 5 cm along bottom) with inner quadrilateral shaded

g circle with radius 12 cm, shaded sector of 30° and another sector of 10°

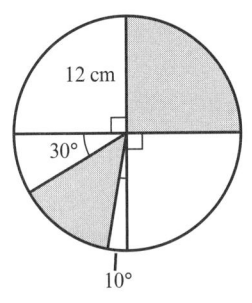

4 $MNOP$ is a trapezium with an area of 150 cm². Calculate the length of NO.

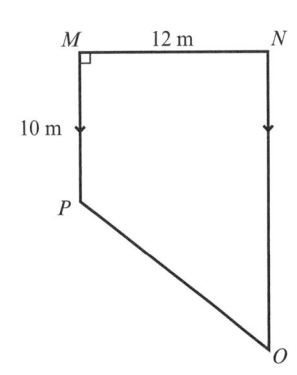

(Diagram: $MN = 12$ m at top, $MP = 10$ m on the left, right angle at M, P at bottom-left, O at bottom-right.)

CONTINUED

5 Calculate the surface area of the pyramid you can make from this net.

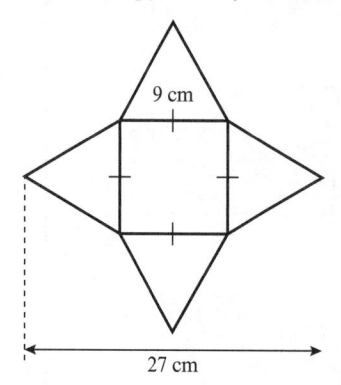

6 Study the two prisms.

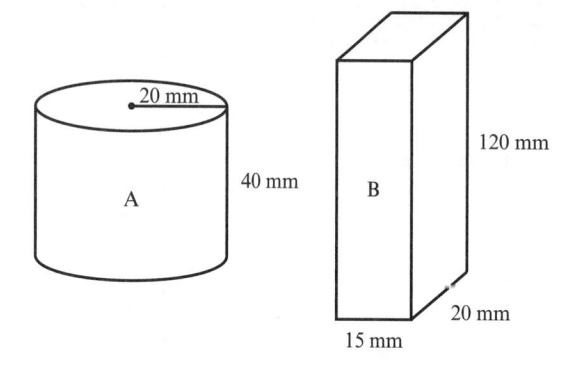

  a Which of the two prisms has the smaller volume?

  b What is the difference in volume? Give your answer correct to two decimal places.

  c Sketch a net of the cuboid. Your net does not need to be to scale, but you must indicate the dimensions of each face on the net.

  d Calculate the surface area of each prism. Give your answer to two decimal places where necessary.

7 How many cubes of side 4 cm can be packed into a wooden box measuring 32 cm by 16 cm by 8 cm?

8 A cylindrical container has a base of area 9π cm² and a height of 20 cm.

  a What is the exact volume of the cylinder?

  b Calculate the volume of the cylinder correct to two decimal places.

  c What is exact circumference of the base of this container?

TIP

When you are asked for an exact answer you give the answer using multiples of π. You can often work out exact answers without using your calculator.

> **CONTINUED**
>
> 9 The photo shows a sphere of ice-cream with a diameter of 6 cm inside a 10 cm long cone of the same diameter.
>
> Calculate the volume of the empty space inside the cone. (Assume that half the sphere of ice-cream is inside the cone although it is no longer spherical!)

Chapter 8: Introduction to probability

8.1 Understanding basic probability

> **KEY LEARNING STATEMENTS**
>
> - Probability is a measure of the chance that something will happen. It is measured on a scale of 0 to 1.
>
> - You can find probabilities through doing an experiment, such as tossing a coin. Each time you perform the experiment is called a trial. If you want to get heads, then heads is your desired (or successful) outcome.
>
> - To calculate probability from the outcomes of experiments, use the formula:
>
> $$\text{Experimental probability of outcome} = \frac{\text{number of successful outcomes}}{\text{number of trials}}$$
>
> - Experimental probability is also called the relative frequency.
>
> - You can calculate the theoretical probability of an event without doing experiments if the outcomes are equally likely. Use the formula:
>
> $$P(\text{outcome}) = \frac{\text{number of favourable outcomes}}{\text{number of possible outcomes}}$$
>
> - You need to work out what *all* the possible outcomes are before you can calculate theoretical probability.
>
> - The probability of an event happening and an event not happening will always add up to one. If A is an event happening, the probability of A not happening = 1 – the probability of A happening.

> **KEY CONCEPTS**
>
> - The probability scale.
> - Calculating the probability of a single event.
> - Calculating the theoretical probability of an event.
> - Calculating the probability that an event does not happen.

1 Salma has a bag containing one red, one white and one green ball. She draws a ball at random and replaces it before drawing again. She repeats this 50 times. She uses a tally table to record the outcomes of her experiment.

Red	|||| |||| ||||
White	|||| |||| |||| |||
Green	|||| |||| |||| ||

  a Calculate the relative frequency of drawing each colour.

  b Express her chance of drawing a red ball as a percentage.

  c What is the sum of the three relative frequencies?

  d What should your chances be in theory of drawing each colour?

2   It is Josh's job to call customers who have had their car serviced at the dealer to check whether they are happy with the service they received. This is the record of what happened for 200 calls made in one month.

Result	Frequency
A: Spoke to customer	122
B: Phone not answered	44
C: Went to voice mail, left message	22
D: Phone engaged or out of order	10
E: Wrong number	2

   a   Calculate the relative frequency of each event as a decimal.

   b   Is it highly likely, likely, unlikely or highly unlikely that the following outcomes will occur when Josh makes a call:

      i   the call will be answered by the customer

      ii the call will go to voice mail

      iii he will have been given a wrong number.

3   This pie chart appeared in an article called 'Stats about cats'. The article reported that there were 10.8 million pet cats in the UK in 2021.

   Number of cats per household

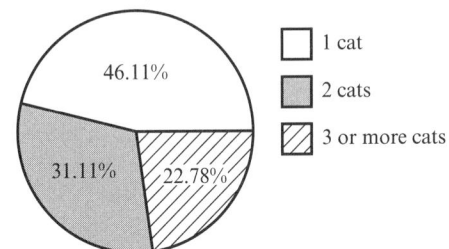

   46.11% — 1 cat
   31.11% — 2 cats
   22.78% — 3 or more cats

   a   What is the probability that one of the 10.8 million pet cats chosen at random is not an only cat?

   b   If a cat-owning household is chosen at random. What is the probability that there are fewer than three cats in the household?

   c   Is it correct to observe that approximately half of the people who own cats have more than one cat? Give a reason for your answer.

4   A container has three red and three blue counters in it. A counter is drawn and its colour is noted.

   a   What are the possible outcomes?

   b   What is the probability of drawing a red counter?

   c   Is it equally likely that you will draw red as blue?

TIP

It is helpful to list the possible outcomes so that you know what to substitute in the formula.

5 An unbiased six-sided dice with the numbers one to six on the faces is rolled.

    a What are the possible outcomes of this event?

    b Calculate the probability of rolling a prime number.

    c What is the probability of rolling an even number?

    d What is the probability of rolling a number greater than seven?

> **TIP**
>
> Remember, 1 is NOT a prime number.

6 Sally has ten identical cards numbered one to ten. She draws a card at random and records the number on it.

    a What are the possible outcomes for this event?

    b Calculate the probability that Sally will draw:

        i the number five   **ii** any one of the ten numbers

        iii a multiple of three   **iv** a number <4

        v a number <5 **vi** a number <6

        vii a square number   **viii** a number <10

        ix a number >10

7 There are five cups of coffee on a tray. Two of them contain sugar.

    a What are your chances of choosing a cup with sugar in it?

    b Which choice is more likely? Why?

8 Mischa has four cards numbered one to four. He draws one card and records the number. Calculate the probability that the result will be:

    a a multiple of three   **b** a multiple of two   **c** factor of three.

9 A dartboard is divided into 20 sectors numbered from 1 to 20. If a dart is equally likely to land in any of these sectors, calculate the probability that the number in the section will be:

    a 8   **b** odd   **c** prime

    d a multiple of 3   **e** a multiple of 5.

10 A school has 40 classrooms numbered from 1 to 40. Work out the probability that a classroom number has the numeral '1' in it.

11 The probability that a driver is speeding on a stretch of road is 0.27. What is the probability that a driver is not speeding?

12 The probability of drawing a green ball in an experiment is $\frac{3}{8}$.

What is the probability of not drawing a green ball?

13 A container holds 300 sweets in five different flavours. The probability of choosing a particular flavour is given in the table.

Flavour	Strawberry	Lime	Lemon	Blackberry	Apple
P(flavour)	0.21	0.22	0.18	0.23	

  a Calculate the probability of choosing an apple flavoured sweet.

  b What is probability of not choosing an apple flavoured sweet?

  c Calculate the probability of choosing neither lemon nor lime flavour.

  d Calculate the number of each flavoured sweet in the packet.

14 In an opinion poll, 5000 teenagers were asked what make of mobile phone they would choose from four options (A, B, C or D). The probability of choosing each option is given in the table.

Phone	A	B	C	D
P(option)	0.36	0.12	0.4	

  a Calculate the missing probability.

  b What is probability of not choosing D?

  c What is the probability a teenager would choose either B or D?

  d How many teenagers would choose option C if these probabilities are correct?

8.2 Sample space diagrams

KEY LEARNING STATEMENTS

- The set of all possible outcomes is called the sample space (or probability space) of an event.
- You use a sample space diagram to show all outcomes clearly.
- When you are dealing with combined events, it is much easier to find a probability if you represent the sample space in a diagram.

KEY CONCEPTS

- Using sample space diagrams to represent all possible outcomes.
- Calculating probability from a sample space diagram.

1 Draw a sample space diagram to show all possible outcomes when you toss two coins at the same time. Use your diagram to help you answer the following.

  a What is the probability of getting at least one tail?

  b What is the probability of getting no tails?

2 Three green cards are numbered one to three and three yellow cards are also numbered one to three.

  a Draw a sample space diagram to show all possible outcomes when one green and one yellow card is chosen at random.

  b How many possible outcomes are there?

c   What is the probability that the number on both the cards will be the same?

d   What is the probability of getting a total <4 if the scores on the cards are added?

3   Two normal six-sided dice are rolled. Find the probability of getting:

a   two fours

b   a four and a six

c   a total of seven

d   a total of nine.

8.3 Combining independent and mutually exclusive events

> **KEY LEARNING STATEMENTS**
>
> - When one outcome in a trial has no effect on the next outcome, the events are independent. For example, drawing a counter at random from a bag, replacing it and then drawing another counter. Because you replace the counter, the first draw does not affect the second draw.
>
> - If A and B *are* independent events, the probability that A happens and then B happens is the product of the probabilities:
>
>   the probability of A and B = the probability of A × the probability of B.
>
> - Mutually exclusive events cannot happen at the same time. For example, you cannot throw an odd number and an even number at the same time when you roll a dice.
>
> - If A and B *are* mutually exclusive events, the probability of A or B happening is the sum of the probabilities:
>
>   the probability of A or B = the probability of A + the probability of B.

> **KEY CONCEPT**
>
> Calculating the probability of simple combined events.

1   Karen enjoys making up games on long bus journeys. One involves choosing a consonant and a vowel at random from the names of places on the journey. The next place is CANBERRA.

a   Draw up a sample space diagram to show all Karen's possible choices.

b   Calculate the probability she chooses R and A.

c   Calculate the probability she chooses R or N and E.

d   Calculate the probability she chooses E and a letter that is not N.

2   Ami has three tiles with the letters A, M and I on them and a coin. If Ami picks a letter at random and tosses the coin, what is the probability of getting the letter M and heads?

3   A box contains four green apples and five red apples. An apple is picked at random, replaced and then a second apple is picked at random. Work out the probability that:

   a   both apples are green

   b   both apples are red

   c   one apple is green and one is red.

REVIEW EXERCISE

1   A coin is tossed a number of times giving the following results.

   Heads: 4083   Tails: 5917

   a   How many times was the coin tossed?

   b   Calculate the relative frequency of each outcome.

   c   What is the probability that the next toss will result in heads?

   d   Jess thinks the results show that the coin is biased. Do you agree? Give a reason for your answer.

2   A bag contains ten red, eight green and two white counters. Each counter has an equal chance of being chosen. Calculate the probability of:

   a   choosing a red counter

   b   choosing a green counter

   c   choosing a white counter

   d   choosing a blue counter

   e   choosing a red or a green counter

   f   not choosing a white counter

   g   choosing a counter that is not red.

3   Two normal unbiased dice are rolled and the sum of the numbers on their faces is recorded.

   a   Calculate the probability the sum is 12.

   b   Which sum has the greatest probability? What is the probability of rolling this sum?

   c   What is the probability of the sum being not even?

   d   What is probability that the sum is <5?

CONTINUED

4 Micah uses a computer to pick a random five-digit positive integer. Work out the probability of choosing:

   a a number ending in nine   **b** an odd number

   c a multiple of five.

5 Grace has two sets of cards. One set contains the letters B, T, L and M, the other contains the letters A, E and O. She picks a card at random from each set and puts them together, consonant first. Calculate the probability that she:

   a makes the word BE or ME   **b** has the letter O in the combination

   c makes the word TEA   **d** does not get a B or an E.

SELF ASSESSMENT

Use this checklist to assess your what you know about basic probability. List the numbers in your book and write 'yes' or 'no' for each statement.

I know that …

1. You can write a number to describe the probability of an event happening.
2. Probability is measured on a scale of 0 to 1.
3. You can represent probability as a fraction, decimal or percentage.
4. You can do experiments to estimate the probability of an event.
5. Experimental probability of outcome = $\dfrac{\text{number of successful outcomes}}{\text{number of trials}}$
6. Experimental probability is also called relative frequency.
7. If outcomes are equally likely you can calculate the theoretical probability of an event by dividing the number of favourable events by the number of possible outcomes.
8. The probability of an event happening and the probability of that event not happening will always sum to 1.
9. Independent events do not affect each other.
10. Mutually exclusive events cannot happen at the same time.
11. Sample space diagrams are a way of showing all possible outcomes using points on a grid.

If you wrote 'no' for any of these points, go back to the relevant work in your coursebook and revise it.

Chapter 9: Sequences and sets

9.1 Sequences

> **KEY LEARNING STATEMENTS**
>
> - A number sequence is a list of numbers that follows a set pattern. Each number in the sequence is called a term. The nth term, or general term is used to describe a term in any position.
> - A linear sequence has a constant difference (d) between the terms. The rule for finding the next term in a sequence is called the term-to-term rule.
> - When you know the rule for making a sequence, you can find the value of any term. Substitute the term number into the rule and solve it.

> **KEY CONCEPTS**
>
> - Describing the rule for continuing a sequence.
> - Finding the nth term.
> - Using the nth term to find the terms in any position.
> - Generating sequences from shape patterns.

> **TIP**
>
> You should recognise these sequences of numbers:
> - square numbers: 1, 4, 9, 16 …
> - cube numbers: 1, 8, 27, 64 …
> - triangular numbers: 1, 3, 6, 10 …
> - Fibonacci numbers: 1, 1, 2, 3, 5, 8 …

1 Find the next three terms in each sequence and describe the term-to-term rule.

    a   11, 13, 15 …

    b   88, 99, 110 …

    c   64, 32, 16 …

    d   8, 16, 24, 32 …

    e   −2, −4, −6, −8 …

    f $\frac{1}{4}, \frac{1}{2}, 1$ …

    g   1, 2, 4, 7 …

    h   1, 6, 11, 16 …

2 List the first four terms of the sequences that follow these rules.

    a   Start with seven and add two each time.

    b   Start with 37 and subtract five each time.

    c   Start with one and multiply by $\frac{1}{2}$ each time.

    d   Start with five then multiply by two and add one each time.

    e   Start with 100 then divide by two and subtract three each time.

3   The rule for finding any term is given for each sequence. In each rule, n represents the term number. List the first three terms of each sequence. Then find the 35th term.

   a $2n + 3$
   b n^2
   c $6n - 1$
   d $n^3 - 1$
   e $n^2 - n$
   f $3 - 2n$

4   Consider the sequence:

   > 2, 10, 18, 26, 34, 42, 50 …

   a   Find the rule for n^{th} term of the sequence.
   b   Find the 200th term.
   c   Which term of this sequence has the value 234? Show full working.
   d   Show that 139 is not a term in the sequence.

5   For each sequence below find the rule for the nth term and work out the 50th term.

   a   7, 9, 11, 13 …
   b   −5, −13, −21, −29 …
   c   2, 8, 14, 20, 26 …
   d   4, 9, 16, 25 …
   e   2.3, 3.5, 4.7, 5.9 …
   f   2, 9, 28, 65 …

REFLECTION

Think about how well you understand the work on sequences.

Answer these questions about your own work:

- Did you work as hard as you could have? What else could you have done?
- How did you review your work to find mistakes and learn from them?
- Was there anything in the exercise that you did not recognise from your work in class? If so, what did you do about it?

9 Sequences and sets

9.2 Rational and irrational numbers

KEY LEARNING STATEMENTS

- You can express any rational number as a fraction in the form $\frac{a}{b}$ where a and b are integers and $b \neq 0$.
- Whole numbers, integers, common fractions, mixed numbers, terminating decimals and recurring decimals are all rational.
- Irrational numbers cannot be written in the form $\frac{a}{b}$. Irrational numbers are all non-recurring, non-terminating decimals.
- The set of real numbers is made up of rational and irrational numbers.

KEY CONCEPT

Identifying and using rational and irrational numbers.

TIP

In $1.\dot{2}$, the dot above the two in the decimal part means it is recurring (the '2' repeats forever). If a set of numbers recurs, e.g. 0.273273273…, there will be a dot at the start and end of the recurring set: $0.\dot{2}7\dot{3}$.

Sometimes a bar is used instead of a dot: $1.\bar{2}$.

1 Write down all the irrational numbers in each set of numbers.

   a $\frac{3}{8}, \sqrt{16}, \sqrt[3]{16}, \frac{22}{7}, \sqrt{12}, 0.090090009\ldots, \frac{31}{3}, 0.020202\ldots$

   b $23, \sqrt{45}, 0.\dot{6}, \frac{3}{4}, \sqrt[3]{90}, \pi, 5\frac{1}{2}, \sqrt{8}, 0.834$

9.3 Sets

KEY LEARNING STATEMENTS

- A set is a list or collection of objects that share a characteristic.
- An element ($\in$) is a member of a set. The number of elements in any set, for example Set A, can be described as n(A).
- A universal set ($\mathscr{E}$) contains all the possible elements appropriate to a particular problem.
- Set A' is the complement of Set A. This is the set of elements that are not in Set A but which are in the universal set ($\mathscr{E}$).
- The elements of two sets can be combined (without repeats) to form the union ($\cup$) of the two sets.
- The elements that two sets have in common is called the intersection ($\cap$) of the two sets.
- A Venn diagram is a pictorial method of showing sets.
- A shorthand way of describing the elements of a set is called set builder notation. For example $\{x : x \text{ is an integer}, 40 < x < 50\}$.

KEY CONCEPTS

- The language and notation used to describe sets.
- Using Venn diagrams to show relationships between sets.

1   Say whether each of the following statements is true or false.

    a $2 \in \{\text{odd numbers}\}$.

    b $8 \in \{\text{cubed numbers}\}$.

    c $\{1, 2, 3\} \cap \{3, 6, 9\} = \{1, 2, 3, 6, 9\}$.

    d $\{1, 2, 3\} \cup \{3, 6, 9\} = \{1, 2, 3, 6, 9\}$.

    e $A = \{1, 2, 3\}, B = \{3, 6, 9\}$, so $A = B$.

2   $\mathscr{E} = \{\text{whole numbers from 1 to 20}\}, A = \{\text{even numbers from 1 to 12}\},$
    $B = \{\text{odd numbers from 1 to 15}\}$ and $C = \{\text{multiples of 3 from 1 to 20}\}$.

    List the elements of the following sets.

    a $A \cap B$ b $B \cup C$

3   List the elements of the following sets.

    a $\{x : x \in \text{integers}, -2 \leq x < 3\}$ b $\{x : x \in \text{natural numbers}, x \leq 5\}$

4   Given Set $A = \{\text{odd numbers between 0 and 10}\}$ and Set $B = \{\text{even numbers between 1 and 9}\}$, find:

    a $A \cup B$ b $A \cap B$

5   List the elements of the following sets.

    a $\{x : x \in \text{integers}, -2 \leq x < 3\}$ b $\{x : x \in \text{natural numbers}, x \leq 5\}$

6   Write in set builder notation.

    a $\{2, 4, 6, 8, 10\}$ b $\{1, 4, 9, 16, 25\}$

7   Draw a Venn diagram to show the following sets and write each element in its correct space.

    $\mathscr{E} = \{\text{letters in the alphabet}\}$

    $P = \{\text{letters in the word physics}\}$

    $C = \{\text{letters in the word chemistry}\}$

8   Use the Venn diagram you drew in question 7 to find:

    a $n(C)$ b $n(\text{not in } P)$ c $C \cap P$ d $P \cup C$

9   $\mathscr{E} = \{\text{whole numbers from 1 to 10}\}$

    $A = \{\text{even numbers from 1 to 10}\}$

    $B = \{\text{multiples of five from 1 to 10}\}$

    a   Draw a Venn diagram to show the information.

    b   Determine:

        i $A \cap B$ ii $n(\text{elements in } \mathscr{E} \text{ but not in } A \text{ or } B)$

        iii $A \cup B$

> **TIP**
> Make sure you know the meaning of the symbols used to describe sets and parts of sets.

> **TIP**
> Sometimes listing the elements of each set will make it easier to answer the question.

> **TIP**
> You can use any shapes to draw a Venn diagram but usually the universal set is drawn as a rectangle and circles within it show the sets.

10 You are told that n(ℰ) = 30, n(A) = 18, n(B) = 12 and n(A ∩ B) = 4.

Draw a Venn diagram to show this information.

11 In a factory that makes T-shirts, 100 T-shirts were quality tested and it was found that 12 had flaws in the printed logos and 15 had flaws in the stitching. 4 T-shirts had both flaws.

  a How many T-shirts had at least one flaw?

  b How many T-shirts had no flaws?

> **TIP**
>
> Venn diagrams are useful for representing the information in problems involving sets of data where some of the data overlaps.

REVIEW EXERCISE

1 For each of the following sequences, find the nth term and the 120th term.

  a 1, 6, 11, 16 …   b 20, 14, 8, 2 …

  c 2, 5, 8, 11 …   d −1, −4, −9, −16 …

2 List the first five terms of the sequence with the rule:

  a $3 - n^2$ b $2n^2 + 3$

3 Here are the first three shapes in a pattern.

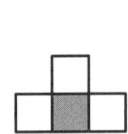

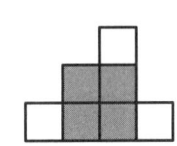

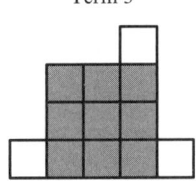

   Term 1   Term 2   Term 3

  a Draw the next shape in the pattern.

  b What is the rule for the nth term in this pattern?

  c How many square will be in the 18th term of this pattern?

4 Which of the following numbers are irrational?

$$1\tfrac{5}{8},\ 0.213231234\ \ldots,\ \sqrt{25},\ \tfrac{7}{17},\ 0.1,\ -0.654,\ \sqrt{2},\ \tfrac{22}{5},\ 4\pi$$

5 Here is a set of numbers:

$\tfrac{2}{3}$	$-\tfrac{3}{5}$	0	−4	25	3.21	$\sqrt{5}$	−2.5	85	0.75

  a Which of these numbers belongs to the set of rational numbers?

  b Which numbers would go into the set A = {integers}?

> **CONTINUED**

6 $\mathscr{E} = \{x : x$ is a whole number between 0 and 11$\}$, 0 and 11 are not elements of $\mathscr{E}$,

$A = \{$even numbers$\}$ and $B = \{$multiples of 3$\}$.

  a Draw a Venn diagram to show the relationships between the sets.

  b List the members of the set $A \cap B$.

  c Determine $n(A \cup B)$.

7 An engineering firm produces metal components for cars. A sample of 120 components was quality tested and it was found that eight had cracks and 11 were not the right size. Three components were both cracked and the incorrect size. Determine the number of components that had:

  a one fault only   **b** no faults   **c** at least one fault.

> Chapter 10: Straight line graphs and quadratic expressions

10.1 Straight line graphs

> **KEY LEARNING STATEMENTS**
>
> - The position of a point can be uniquely described on the coordinate grid using ordered pairs (x, y) of coordinates.
>
> - You can use equations in terms of x and y to generate a table of paired values for x and y. You can plot these on the coordinate grid and join them to draw a graph. To find y-values in a table of values, substitute the given (or chosen) x-values into the equation and solve for y.
>
> - The gradient of a line describes its slope or steepness. Gradient can be defined as:
>
> $$m = \frac{\text{change in } y}{\text{change in } x}$$
>
>   - lines that slope up to the right have a positive gradient
>   - lines that slope down to the right have a negative gradient
>   - lines parallel to the x-axis (horizontal lines) have a gradient of 0
>   - lines parallel to the y-axis (vertical lines) have an undefined gradient
>   - lines parallel to each other have the same gradients.
>
> - You can write the equation of a straight line in general terms as $y = mx + c$, where x and y are coordinates of points on the line, m is the gradient of the line and c is the y-intercept (the point where the graph crosses the y-axis).
>
> - To find the equation of a given line you need to find the y-intercept and substitute this for c. Then you need to find the gradient of the line and substitute this for m.

> **KEY CONCEPTS**
>
> - Plotting straight line graphs and finding the equation of a graph.
> - Calculating the gradient of a line and lines parallel to it.

1 For x-values of $-1, 0, 1, 2$ and 3, draw a table of values for each of the following equations.

   a $y = x + 5$ b $y = -2x - 1$ c $y = 7 - 2x$ d $y = -x - 2$
   e $x = 4$ f $y = -2$ g $y = -2x - \frac{1}{2}$ h $4 = 2x - 5y$
   i $0 = x - 2y - 1$ j $x + y = -\frac{1}{2}$

> **TIP**
>
> Normally the x-values will be given. If not, choose at least three small values (for example, $-2, 0$ and 2). All graphs should be clearly labelled with their equation.

2   Draw and label graphs (a) to (e) in question 1 on one set of axes and graphs (f) to (j) on another.

3   Find the equation of a line parallel to $y = x + 5$ and passing through point $(0, -2)$.

4   Are the following pairs of lines parallel?

   a $y = 3x + 3$ and $y = x + 3$

   b $y = \frac{1}{2}x - 4$ and $y = \frac{1}{2}x - 8$

   c $y = -3x$ and $y = -3x + 7$

   d $y = 0.8x - 7$ and $y = 8x + 2$

   e $2y = -3x + 2$ and $y = \frac{3}{2}x + 2$

   f $2y - 3x = 2$ and $y = -1.5x + 2$

   g $y = 8$ and $y = -9$

   h $x = -3$ and $x = \frac{1}{2}$

> **TIP**
>
> Remember, parallel lines have the same gradient.

5   Find the gradient of each of the following lines.

a

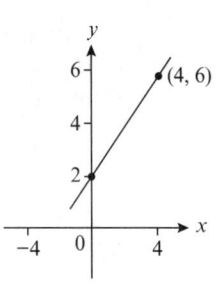

b

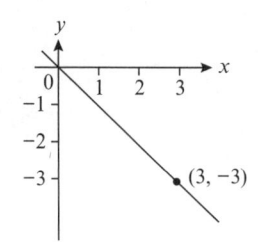

c

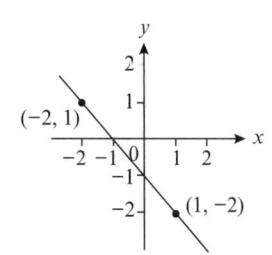

d

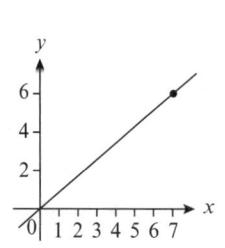

e

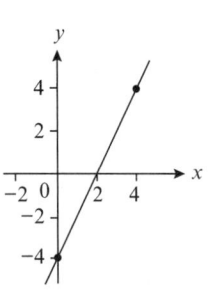

f

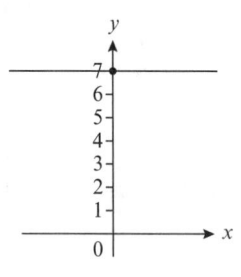

g

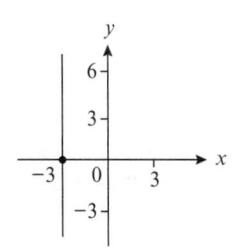

h

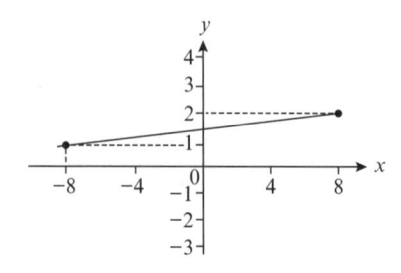

6 Write down the gradient (*m*) and the *y*-intercept (*c*) of each of the following graphs.

    a $y = 3x - 4$ **b** $y = -x - 1$ **c** $y = -\frac{1}{2}x + 5$ **d** $y = x$

    e $y = \frac{x}{2} + \frac{1}{4}$ **f** $y = \frac{4x}{5} - 2$ **g** $y = 7$ **h** $y = -3x$

7 Determine the equation of each of the following graphs.

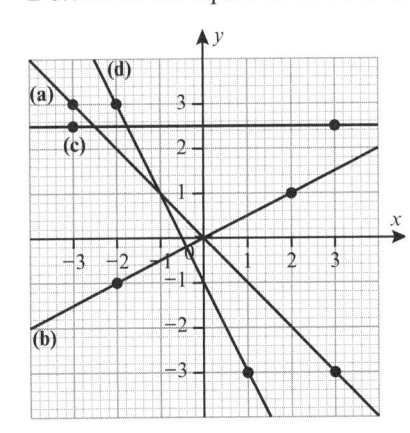

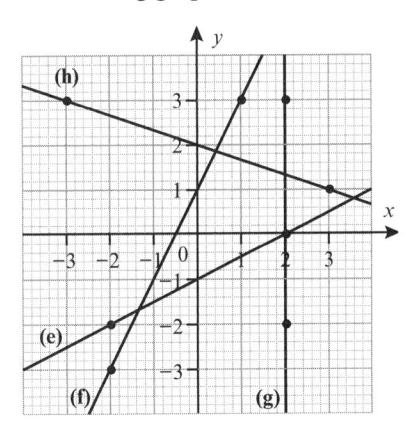

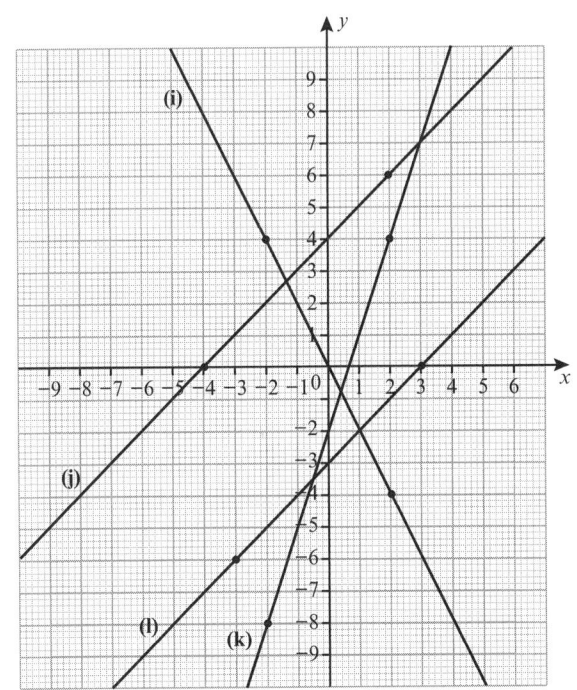

8 Find the *x*- and *y*-intercepts of the following lines.

    a $y = 3x - 6$ **b** $y = -\frac{1}{2}x + 3$ **c** $2y - 3x = 12$

    d $\frac{x + y}{2} = 5$ **e** $2x + y + 5 = 0$

10.2 Quadratic expressions

> **KEY LEARNING STATEMENTS**
>
> - A quadratic expression has terms where the highest power of the variable is two (for example x^2).
> - You can expand (multiply out) the product of two brackets by multiplying each term of the first bracket by each term of the second. You may then need to add or subtract any like terms.

KEY CONCEPT

Expanding products of two sets of brackets.

1   Expand and simplify.

   a $(x + 2)(x + 3)$
   b $(x + 2)(x - 3)$
   c $(x + 5)(x + 7)$
   d $(x - 5)(x + 7)$
   e $(x - 1)(x - 3)$
   f $(2x - 1)(x + 1)$
   g $(y - 7)(y - 2)$
   h $(2x - y)(3x - 2y)$
   i $(x^2 + 1)(2x^2 - 3)$
   j $(x - 11)(x + 12)$
   k $\left(\frac{1}{2}x + 1\right)\left(1 - \frac{1}{2}x\right)$
   l $(x - 3)(2 - 3x)$
   m $(3x - 2)(2 - 4x)$

TIP

The acronym, FOIL, may help you to systematically expand pairs of brackets:

F – first × first
O – outer × outer
I – inner × inner
L – last × last

2   Expand and simplify.

   a $(x + 4)^2$
   b $(x - 3)^2$
   c $(x + 5)^2$
   d $(y - 2)^2$
   e $(x + y)^2$
   f $(2x - y)^2$

3   A carpet fitter has a square of carpet with sides x metres long. To fit the room he needs to cut a 40 cm wide strip off the edge of the square piece and place it along the adjacent side of the square forming a rectangle. Any carpet not needed will be discarded.

   a   Express the length and width of the rectangular carpet in terms of x.
   b   Write an expression for the area of the rectangular piece.
   c   What area of carpet is discarded?

TIP

Drawing a diagram will help to make this problem clearer.

10 Straight line graphs and quadratic expressions

SELF ASSESSMENT

1. Think about what you need to know and remember to expand expressions in brackets. Develop a list of three criteria that you can use to decide whether you have met the learning intentions.

2. Once you have your list, check your work against the success criteria.
   - Place a tick (✓) next to a criterion if you can find evidence that you have met it. (You are looking for how your work shows you have achieved the criteria.)
   - If you cannot find evidence, write down what you can do to improve in that particular area.

3. Look at the success criteria that are not present in your work. Ask yourself why these are not present.
   - Are you finding that success criteria challenging? If so, you could discuss this with a partner, ask your teacher for help or find an online lesson or tutorial on this topic.

4. Make any improvements that you need to over the next few days and then reassess your work using the same criteria.

REVIEW EXERCISE

1. For each equation, copy and complete the table of values. Draw the graphs for all four equations on the same set of axes.

   a $y = \frac{1}{2}x$

x	−1	0	1	2	3
y					

   b $y = \frac{1}{2}x + 3$

x	−1	0	1	2	3
y					

   c $y = 2$

x	−1	0	1	2	3
y					

   d $y - 2x - 4 = 0$

x	−1	0	1	2	3
y					

CONTINUED

2 Determine the gradient and the y-intercept of each graph.

    a $y = -2x - 1$ **b** $y = x - 6$

    c $y = -\dfrac{1}{2}$ **d** $y = -x$

3 What equation defines each of these lines?

    a A line with a gradient of 1 and a y-intercept of -3.

    b A line with a y-intercept of $\dfrac{1}{2}$ and a gradient of $-\dfrac{2}{3}$.

    c A line parallel to $y = -x + 8$ with a y-intercept of -2.

    d A line parallel to $y = -\dfrac{4}{5}x$ which passes through the point $(0, -3)$.

    e A line parallel to $2y - 4x + 2 = 0$ with a y-intercept of -3.

    f A line parallel to $x + y = 5$ which passes through $(1, 1)$.

    g A line parallel to the x-axis which passes through $(1, 2)$.

    h A line parallel to the y-axis which passes through $(-4, -5)$.

4 Find the gradient of the following lines.

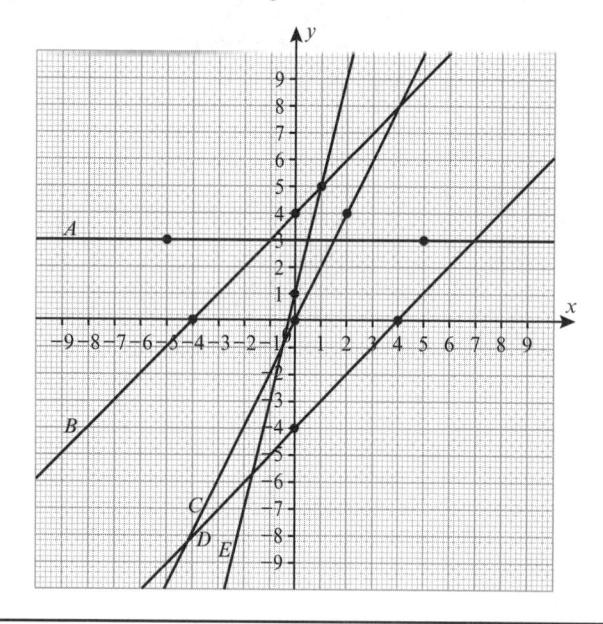

> **TIP**
>
> The equation that defines the line is the same as the equation of the line.

CONTINUED

5 What is the equation of each of these lines?

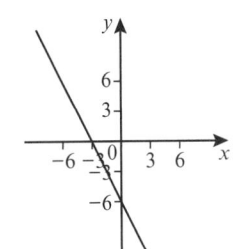

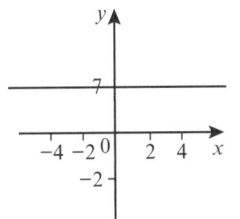

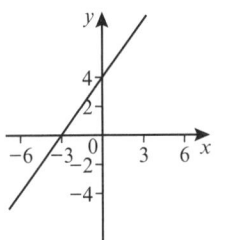

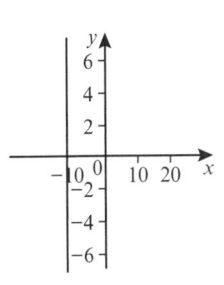

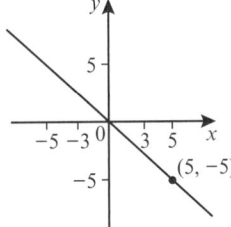

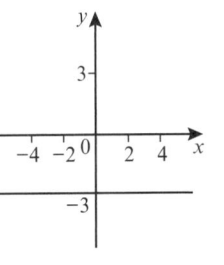

a, b, c, d, e, f (graphs)

6 Caroline likes running. She averages a speed of 7 km/h when she runs. This relationship can be expressed as $D = 7t$, where D is the distance covered and t is the time (in hours) that she runs for.

  a Use the formula $D = 7t$ to make a table of values for 0, 2, 4 and 6 hours of running.

  b On a set of axes, draw a graph to show the relationship between D and t. Think carefully about how you will number the axes before you start.

  c Write an equation in the form of $y = mx + c$ to describe this graph.

  d What is the gradient of the line?

  e Use your graph to find the time it takes Caroline to run:

   i 21 km   ii 10 km   iii 5 km.

  f Use your graph to find out how far she runs in:

   i 3 hours   ii $2\frac{1}{2}$ hours   iii $\frac{3}{4}$ of an hour.

7 Expand and simplify.

  a $(x + 12)(x - 2)$ b $(x - 8)(x + 5)$

  c $(2x + 5)(2x + 4)$ d $(2x + 3)^2$

8 Work out the missing values in each equation.

  a $(x + 6)(x + 8) = x^2 + \Box x + \Box$

  b $(x + \Box)(x + 6) = x^2 + 10x + \Box$

  c $(\Box x + \Box)(2x + 7) = 4x^2 + 18x + \Box$

> **TIP**
>
> Time is usually plotted on the horizontal or x-axis because it is the independent variable in most relationships. In this graph you will only need to work in the first quadrant. You won't have any negative values because Caroline cannot run for less than 0 hours and her speed cannot be less than 0 km/h.

Chapter 11: Pythagoras' theorem and similar shapes

11.1 Pythagoras' theorem

KEY LEARNING STATEMENTS

- In a right-angled triangle, the square of the length of the hypotenuse (the longest side) is equal to the sum of the squares of the lengths of the other two sides. This can be expressed as $c^2 = a^2 + b^2$, where c is the hypotenuse and a and b are the two shorter sides of the triangle.

- Conversely, If $c^2 = a^2 + b^2$ then the triangle is right angled.

- To find the length of an unknown side in a right-angled triangle you need to know two of the sides. Then you can substitute the two known lengths into the formula and solve for the unknown length.

KEY CONCEPTS

- Applying Pythagoras' theorem to calculate side lengths in a right-angled triangle.

- Using Pythagoras' theorem to determine whether a triangle is right angled.

TIP

The hypotenuse is the longest side. It is always opposite the right angle.

1 Calculate the length of the unknown side in each of these triangles.

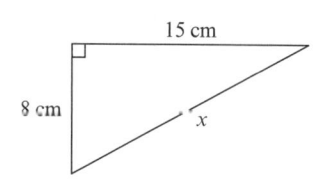

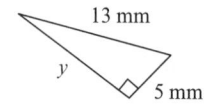

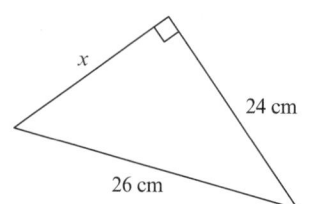

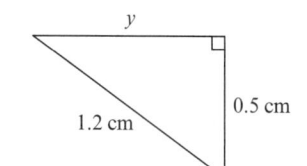

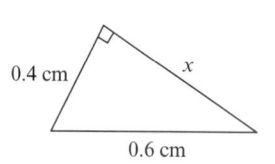

a 3 cm, 4 cm, x

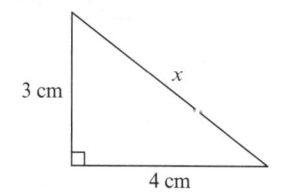

b 15 cm, 8 cm, x

c 13 mm, 5 mm, y

d x, 24 cm, 26 cm

e y, 1.2 cm, 0.5 cm

f 0.4 cm, 0.6 cm, x

g x, 11 cm, 7 cm

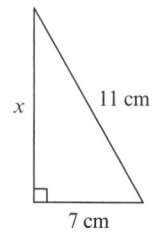

h y, 4 cm, 7.3 cm

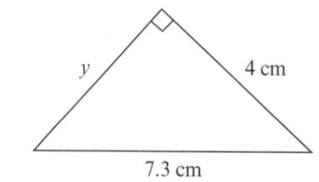

2 Find the length of the side marked with a letter in each figure.

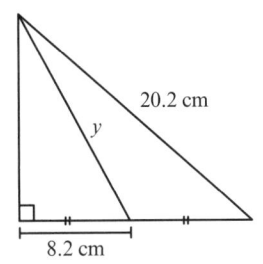

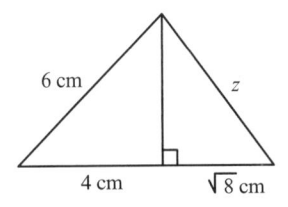

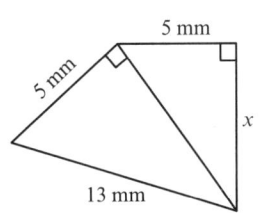

a [Figure: right triangle with legs 80 mm and 70 mm, and another right triangle sharing the hypotenuse with base 120 mm and side x]

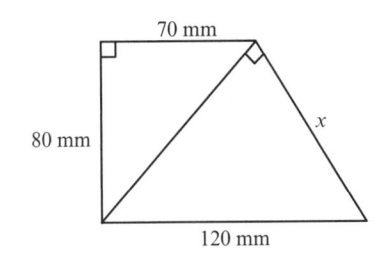

b [Figure: triangle with hypotenuse 20.2 cm, internal segment y, and base 8.2 cm marked with tick marks]

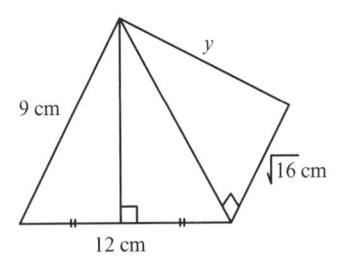

c [Figure: triangle with sides 6 cm and z, with perpendicular from apex to base splitting base into 4 cm and $\sqrt{8}$ cm]

d [Figure: quadrilateral split by diagonal x; one right triangle has legs 5 mm and 5 mm, the other has side 13 mm]

e [Figure: triangle with side 9 cm, apex to side y, perpendicular from apex to base 12 cm, with $\sqrt{16}$ cm segment]

f [Figure: triangle with sides 9 cm, 5 cm on left; 15 cm on upper right; base 16 cm; internal segment z]

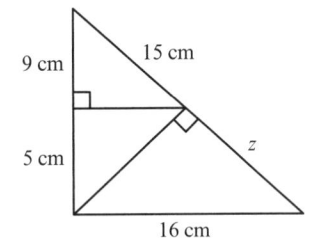

3 Determine whether each of the following triangles is right angled. Side lengths are all in centimetres.

a [Triangle ABC with $AB = 7$, $AC = 8$, $BC = 10$]

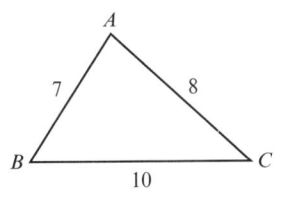

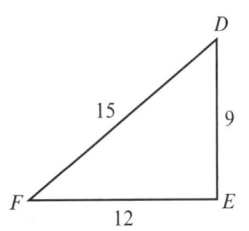

b [Triangle DEF with $DF = 15$, $DE = 9$, $FE = 12$]

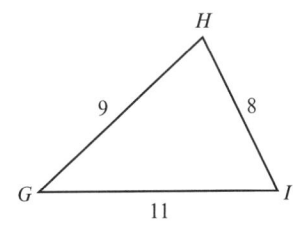

c [Triangle GHI with $GH = 9$, $HI = 8$, $GI = 11$]

d [Triangle JKL with $JK = 12$, $JL = 13$, $KL = 5$]

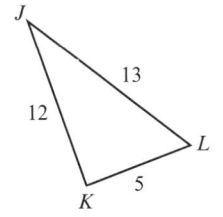

4 A rectangle has sides of 12 mm and 16 mm. Calculate the length of one of the diagonals.

5 The size of a rectangular computer screen is determined by the length of the diagonal. Nick buys a 55 cm screen that is 33 cm high. How long is the base of the screen?

6 The sides of an equilateral triangle are 100 mm long. Calculate the perpendicular height of the triangle and hence find its area.

TIP

With word problems, make a rough sketch of the situation, fill in the known lengths and mark the unknown lengths with letters.

CAMBRIDGE IGCSE™ MATHEMATICS: CORE PRACTICE BOOK

7   A vertical pole is 12 metres long. It is supported by two ropes that are attached to the top of the pole and fixed to the ground. One rope is fixed to the ground 5 metres from the base of the pole and the other rope is fixed to the ground 9 metres from the base of the pole. Calculate the length of each rope.

8   Geetha flew a drone 2000 metres on a bearing of 090°, then 1000 metres due north and then 800 metres on a bearing of 270°. Sketch the journey of the drone and then calculate the straight line distance from the starting point to the finishing point.

11.2 Understanding similar triangles

KEY LEARNING STATEMENTS

- Triangles are similar when the corresponding sides are proportional and the corresponding angles are equal in size.
- In similar shapes the ratios of corresponding sides are all equal, so dividing the length of a side by the length of its corresponding side will always give you the same answer.

KEY CONCEPT

Calculating lengths of sides in similar triangles.

1   Write down three pairs of triangles that are similar.

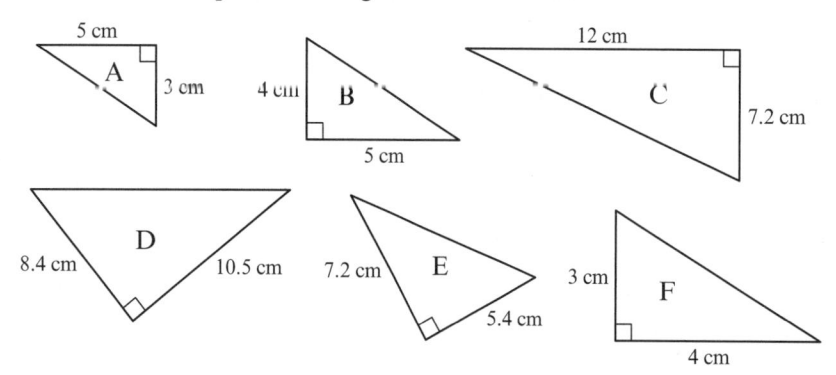

Triangle A: 5 cm, 3 cm
Triangle B: 4 cm, 5 cm
Triangle C: 12 cm, 7.2 cm
Triangle D: 8.4 cm, 10.5 cm
Triangle E: 7.2 cm, 5.4 cm
Triangle F: 3 cm, 4 cm

TIP

Work out which sides are corresponding before you start. It is helpful to mark corresponding sides in the same colour or with a symbol.

2   The pairs of triangles in this question are similar. Calculate the unknown (lettered) length in each case.

a   Triangle 1: 4.47 cm, 2 cm, 4 cm. Triangle 2: x, 2 cm, 1 cm.

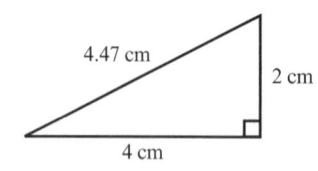

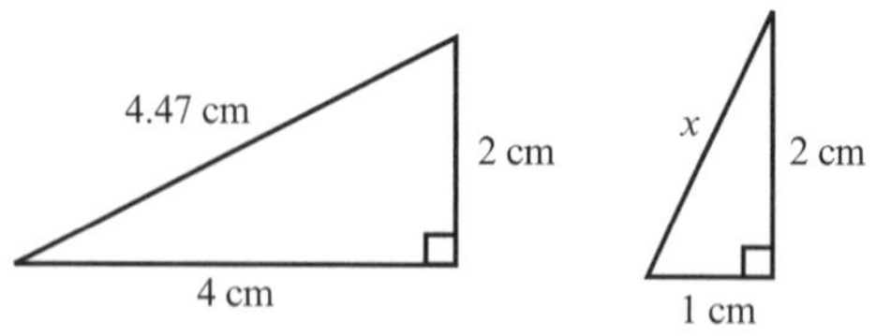

b   Triangle 1: 12 mm, 6 mm, 9 mm, with angles α and β. Triangle 2: 8 mm, 4 mm, y, with angles α and β.

88

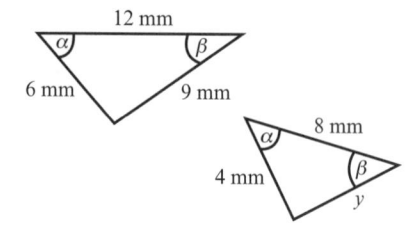

11 Pythagoras' theorem and similar shapes

c
[Triangle with sides 6 mm, 3 mm, side y, angles α and β]
[Triangle with 5 mm, 2 mm, 4 mm, angles α and β]

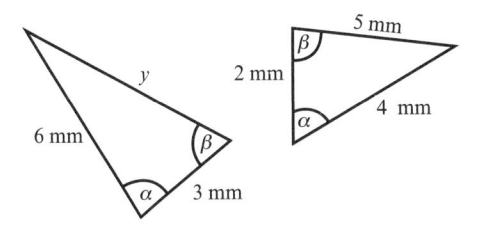

d
[Triangle with 8 cm, 10 cm, angles x and y]
[Triangle with 8 cm, angles x, y, z]

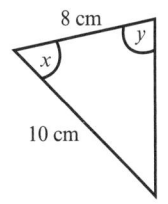

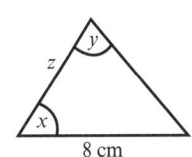

e
[Triangle with 8 cm, 10 cm, side y, angles α and β]
[Triangle with 4 cm, 6 cm, side z, angles α and β]

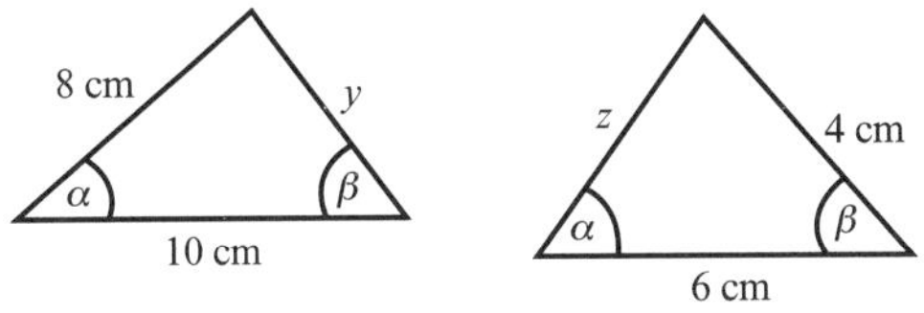

f
[Triangle with 8 cm, 5 cm, side x, angle 43°, angle α]
[Triangle with 8.5 cm, 9.5 cm, 43°, angle α, side y]

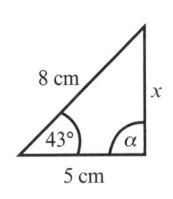

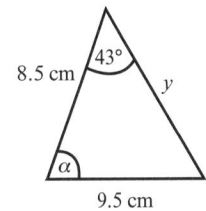

g
[Triangle with 8 cm, 7 cm, side x, angles α and β]
[Triangle with 27 cm, 21 cm, side y, angles α and β]

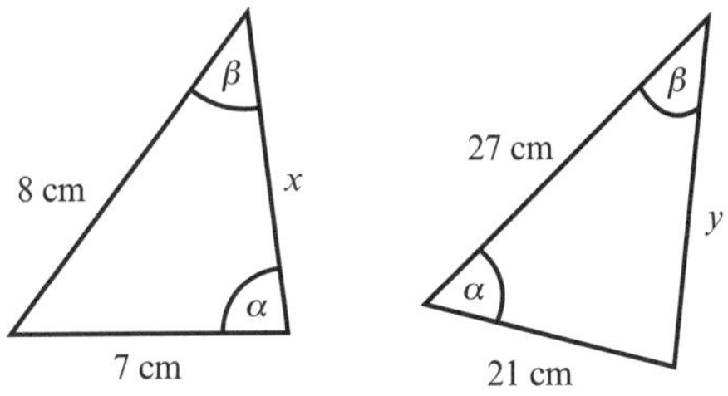

h
[Triangle with 40 cm, 30 cm, side x, angles α and β]
[Triangle with 25 cm, 15 cm, side y, angles α and β]

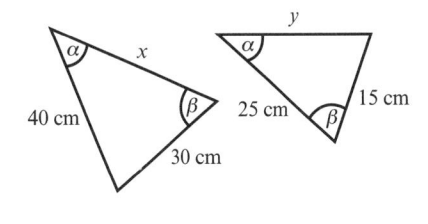

3 Explain fully why triangle ABC is similar to triangle ADE.

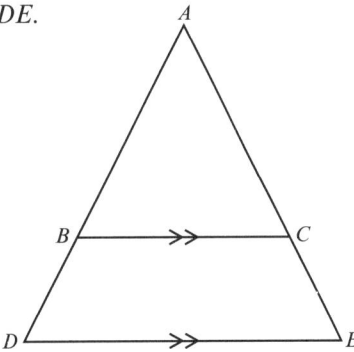

4 Nancy is lying on a blanket on the ground, 4 metres away from a 3 metre tall tree. When she looks up past the tree she can see the roof of a building which is 30 metres beyond the tree. Work out the height of the building.

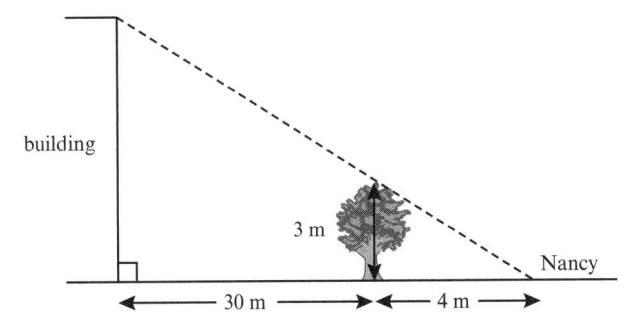

89

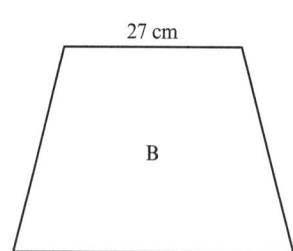

11.3 Understanding similar shapes

KEY LEARNING STATEMENTS

- The ratio of corresponding sides in similar shapes is equal. The lengths of unknown sides can be found by the same method used for similar triangles.

KEY CONCEPT

Calculating lengths in similar shapes.

1   Miri wants to make four similar rectangular tiles. She works out these dimensions:

   A   120 × 80 mm

   B   90 × 60 mm

   C   180 × 120 mm

   D   240 × 200 mm

   a   Which dimensions are incorrect?

   b   She makes another tile similar to tile A that is 150 mm long. How wide will it be?

2   Find the length of each side marked with a letter in these pairs of similar shapes. All dimensions are in centimetres.

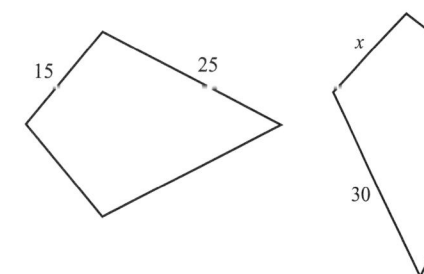

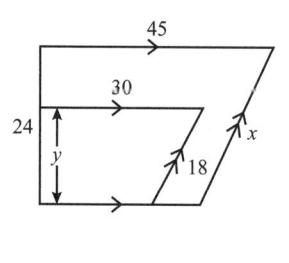

3   These two quadrilaterals are similar. Show how you can use the given information to work out the perimeter of shape B without calculating the lengths of the unmarked sides.

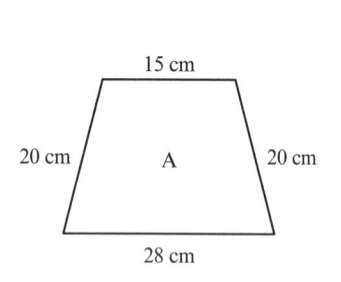

11.4 Understanding congruence

KEY LEARNING STATEMENTS

- Congruent shapes are identical in shape and size.

KEY CONCEPT

Recognising congruent shapes.

1   Identify all the shapes that are congruent to *A*.

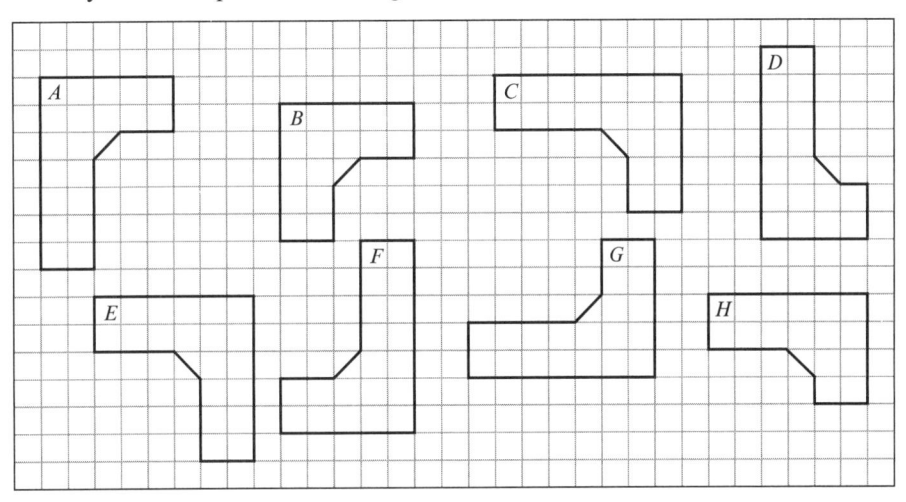

2   Use squared or dotted paper to investigate how many triangles, congruent to this one you can draw on a 3 × 3 grid.

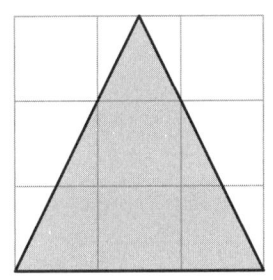

3   The two right-angled triangles in the figure below each have a side 3 cm long and an angle of 30°.

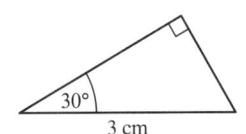

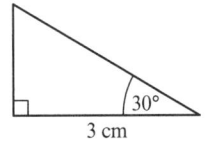

   a   Do the two triangles contain three pairs of equal angles? Give a reason for your answer.

   b   Are these two triangles congruent? How did you decide?

   c   Draw a diagram to show that these two triangles could be similar.

4   The two triangles in this diagram are congruent.

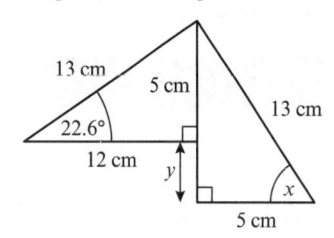

    a   What is the size of x?   b   What is the length y?

5   In the imperial system, 1 foot = 12 inches. An inch is equivalent to 2.54 cm.

    a   Line MN is 2.5 foot long and Line PQ is 2 foot 6 inches long.
        Are the lines congruent?

    b   Line RS is congruent to MN. What is the length of line RS in centimetres?

REVIEW EXERCISE

1   A school caretaker wants to mark out a sports field 50 metres wide and 120 metres long. To make sure that the field is rectangular, he needs to know how long each diagonal should be.

    a   Draw a rough sketch of the field.

    b   Calculate the required lengths of the diagonals.

2   In triangle ABC, $AB = 10$ cm, $BC = 8$ cm and $AC = 6$ cm. Determine whether the triangle is right angled or not and give reasons for your answer.

3   A triangle with sides of 25 mm, 65 mm and 60 mm is similar to another triangle with its longest side 975 mm. Calculate the perimeter of the larger triangle.

4   Calculate the missing dimensions in each of these pairs of similar triangles.

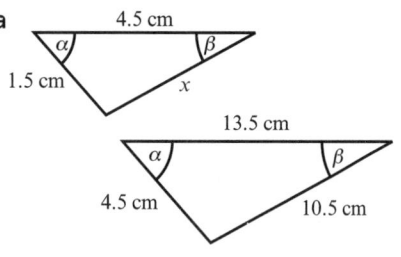

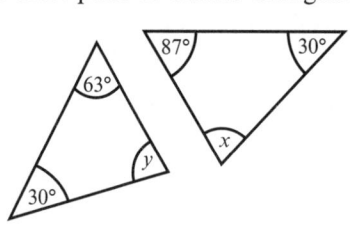

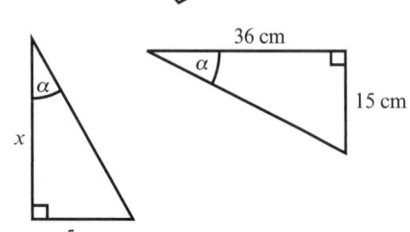

CONTINUED

5 Which triangles are congruent in this set?

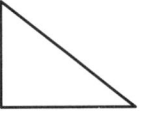

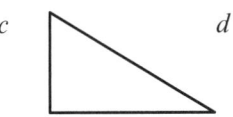

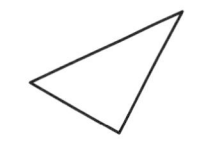

a b c d

6 Read each statement. Decide whether it is true or mathematically false. Give a reason for each choice.

  a If two rectangles are congruent, their perimeters are equal.

  b If triangle A is congruent to triangle B, the triangles will have the same area.

  c If two triangles have the same area they will be congruent.

  d If two quadrilaterals have a perimeter of 24 cm they will be congruent.

7 An 8.6 metre long wire cable is used to secure a mast of height x metres. The cable is attached to the top of the mast and secured on the ground 6.5 metres away from the base of the mast. How tall is the mast? Give your answer correct to two decimal places.

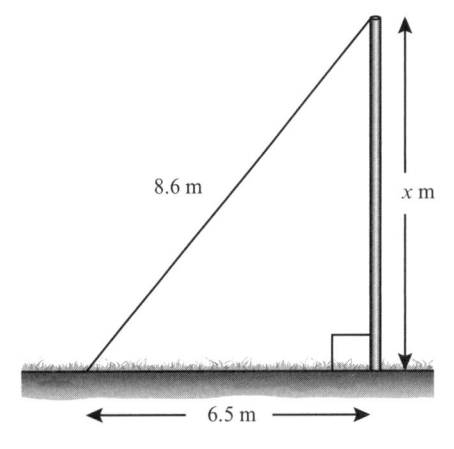

8 Nadia wants to have a metal number 1 made for her gate. She has found a sample brass numeral and noted its dimensions. She decides that her numeral should be similar to this one, but that it should be four times larger.

  a Draw a rough sketch of the numeral that Nadia wants to make with the correct dimensions written on it in millimetres.

  b Calculate the length of the sloping edge at the top of the full size numeral to the nearest whole millimetre.

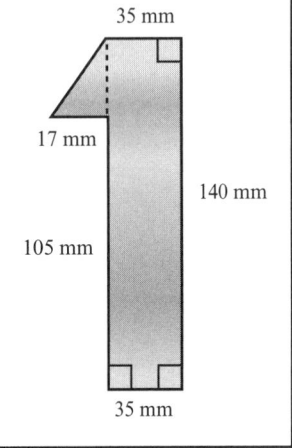

SELF ASSESSMENT

1. Check your own work and answers to the Review exercise.

2. Write the numbers 1 to 7 in your book. Use the following scale and symbols to rate your understanding and ability for each question.

   No competence
   Low competence
   Some competence
   High competence
   Expert

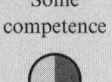

3. Write down three steps you will take to improve any areas where you have rated yourself ◯, ◔ or ◑.

Chapter 12: Averages and measures of spread

12.1 Different types of average

> **KEY LEARNING STATEMENTS**
> - An average is a measure of central tendency in a data set.
> - There are three main types of average: mean, median and mode.
> - The mean is the sum of the data items divided by the number of items in the data set. The mean does not have to be one of the numbers in the data set.
>   - The mean can be affected by extreme values in the data set. When one value is much lower or higher than the rest of the data it is called an outlier. Outliers skew the mean and make it less representative of the data set.
> - The median is the middle value in a set of data when the data is arranged in order.
>   - When there is an even number of values, the median is the mean of the two middle values.
> - The mode is the number (or item) that appears most often in a data set.
>   - When two numbers appear most often the data has two modes. When more than two numbers appear equally often the mode has no real value as a statistic.

> **KEY CONCEPT**
> Calculating different types of averages (mean, median and mode) for individual discrete data.

1. Determine the mean, median and mode of the following sets of data.

   a 5, 9, 6, 4, 7, 6, 6

   b 23, 38, 15, 27, 18, 38, 21, 40, 27

   c 12, 13, 14, 12, 12, 13, 15, 16, 14, 13, 12, 11

   d 4, 4, 4, 5, 5, 5, 6, 6, 6

   e 4, 4, 4, 4, 5, 5, 6, 6, 6

   f 4, 4, 5, 5, 5, 6, 6, 6, 6

2. Five students scored a mean mark of 14.8 out of 20 for a maths test.

   a Which of these sets of marks fit this average?

       A 14, 16, 17, 15, 17   B 12, 13, 12, 19, 19   C 12, 19, 12, 18, 13

       D 13, 17, 15, 16, 17   E 19, 19, 12, 0, 19   F 15, 15, 15, 15, 14

   b Compare the sets of numbers in your answer to part (a). Explain why you can get the same mean from different sets of numbers.

> **TIP**
> If you multiply the mean by the number of items in the data set, you get the total of the scores. This will help you solve problems like question 2.

3 The mean of 15 numbers is 17. What is the sum of the numbers?

4 The sum of 21 numbers is 312.8. Which of the following numbers is closest to the mean of the 21 numbers? 14, 15, 16 or 17.

5 The heights (to the nearest metre) of different varieties of palm trees in a botanical garden are shown on the chart.

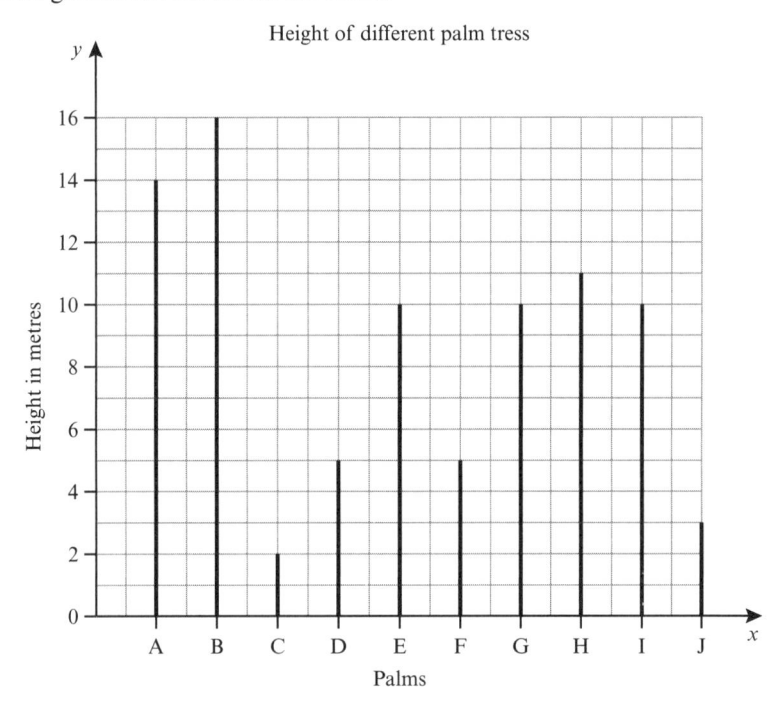

Height of different palm tress

- a What is the range of heights?
- b What is the mean height?
- c The windmill palm tree is the tree with the median height in this data set. How tall is it?
- d What is the most common height of tree in this data set?

6 An agricultural worker wants to know which of two dairy farms have the best milk-producing cows. The cows on farm A produce 2490 litres of milk per day. The cows on farm B produce 1890 litres of milk per day. There is not enough information to decide which cows are the better producers of milk. What other information do you need to answer the question?

7 In a group of students, six had four siblings, seven had five siblings, eight had three siblings, nine had two siblings and ten had one sibling. (Siblings are brothers and sisters.)

- a What is the total number students?
- b What is the total number of siblings?
- c What is the mean number of siblings?
- d What is the modal number of siblings?

> **TIP**
>
> It may help to draw up a rough frequency table to solve problems like this one.

8 The management of a factory announced salary increases and said that workers would receive an average increase of $20 to $40.

The table shows the old and new salaries of the workers in the factory.

	Previous salary	Salary with increase
Four workers in Category A	$180	$240
Two workers in Category B	$170	$200
Six workers in Category C	$160	$170
Eight workers in Category D	$150	$156

a Calculate the mean increase for all workers.

b Calculate the modal increase.

c What is the median increase?

d How many workers received an increase of between $20 and $40?

e Was the management announcement true? Say why or why not.

9 This stem-and-leaf diagram shows the number of text messages that Mario received on his phone each day for a month.

Number of text messages

Stem	Leaf
2	3 5 5 6 6 6 7
3	2 4 4 6 8 8 8 8 9 9
4	0 1 2 4 4 4 5 7 8 8
5	0 1 5

Key: 2|3 represents 23 messages

a What is the range of the data?

b What is the mode?

c Determine the median number of messages received.

12.2 Making comparisons using averages and ranges

> **KEY LEARNING STATEMENTS**
>
> - You can use averages to compare two or more sets of data. However, averages on their own may be misleading, so it is useful to work with other summary statistics as well.
> - The range is a measure of how spread out (dispersed) the data is. Range = largest value − smallest value.
> - A large range means that the data is spread out, so an average may not be representative of the whole data set.

> **KEY CONCEPTS**
>
> - Calculating the range.
> - Using statistical measures to compare sets of data.

1   For each of these sets of data, one of the three averages is not representative. State which one is not representative in each case.

   a   6, 2, 5, 1, 5, 7, 2, 3, 8

   b   2, 0, 1, 3, 1, 6, 2, 9, 10, 3, 2, 2, 0

   c   21, 29, 30, 14, 5, 16, 3, 24, 17

> **TIP**
>
> When the mean is affected by extreme values the median is more representative of the data.

2   Twenty students scored the following results in a test (out of 20).

| 17 | 18 | 17 | 14 | 8 | 3 | 15 | 18 | 3 | 15 |
| 0 | 17 | 16 | 17 | 14 | 7 | 18 | 19 | 5 | 15 |

   a   Calculate the mean, median, mode and range of the marks.

   b   Why is the median the best summary statistic for this particular set of data?

> **TIP**
>
> The mode only tells you the most popular value and this is not necessarily representative of the whole data set.

3   The table shows the times (in minutes and seconds) that two 800 metre runners achieved during one season.

| Runner A | 2 min 2.5 s | 2 min 1.7 s | 2 min 2.2 s | 2 min 3.7 s | 2 min 1.7 s | 2 min 2.9 s | 2 min 2.6 s |
| Runner B | 2 min 2.4 s | 2 min 1.8 s | 2 min 2.3 s | 2 min 4.4 s | 2 min 0.6 s | 2 min 2.2 s | 2 min 1.2 s |

   a   Which runner is better? Why?

   b   Which runner is more consistent? Why?

4   Ten people were selected randomly and asked how long they waited to see a consultant at the passport office. The waiting times in minutes were:

| 16 | 25 | 30 | 112 | 20 | 10 | 17 | 40 | 22 | 10 |

Which average best describes this data? Give two reasons for your choice.

12.3 Calculating averages and ranges for frequency data

KEY LEARNING STATEMENTS

- The mean can be calculated from a frequency table. To calculate the mean you add a column to the table and calculate the score × frequency (fx).

   Mean = $\dfrac{\text{total of (score} \times \text{frequency) column}}{\text{total of frequency column}}$.

- Find the mode in a table by looking at the frequency column. The data item with the highest frequency is the mode.

- In a frequency table, the data is already ordered by size. To find the median, work out its position in the data and then add the frequencies until you equal or exceed this value. The score in this category will be the median.

- You can find the mean, median, mode and range for a data set using an ordered stem-and-leaf diagram. The diagram also shows the distribution of the data visually.

KEY CONCEPTS

- Calculating averages when data is in a frequency table or diagram.
- Interpreting frequency tables and stem-and-leaf diagrams.

1 Copy and complete the frequency table for the data below and then calculate:

   a the mean   b the mode   c the median   d the range.

0	3	4	3	3	2	2	2	2	1
3	3	4	3	6	2	2	2	0	0
5	4	3	2	4	3	3	3	2	1
3	1	1	1	1	0	0	0	2	4

Score	Frequency
0	
1	
2	
3	
4	
5	
6	

2 For each of the following frequency distributions calculate:

   a the mean score   b the median score   c the modal score.

   Data set A

Score	1	2	3	4	5	6
Frequency	12	14	15	12	15	12

   Data set B

Score	10	20	30	40	50	60	70	80
Frequency	13	25	22	31	16	23	27	19

   Data set C

Score	1.5	2.5	3.5	4.5	5.5	6.5
Frequency	15	12	15	12	10	21

3   The ages of children playing in the ball pit at a fast-food restaurant during one afternoon were recorded and tabulated.

Age	2	3	4	5	6	7	8	9	10
Frequency	1	10	11	10	8	5	4	3	2

   a   What is the range of ages?

   b   What is the modal age?

   c   What is the mean age?

   d   What is the median age?

REVIEW EXERCISE

1   Find the mean, median, mode and range of the following sets of data.

   a   6   5   6   7   4   5   8   6   7   10

   b   6   3   2   4   2   1   2   2   1

   c   12.5   13.2   19.4   12.8   7.5   18.6   12.6

2   The mean of two consecutive numbers is 9.5. The mean of eight different numbers is 4.7.

   a   Calculate the total of the first two numbers.

   b   What are these two numbers?

   c   Calculate the mean of the ten numbers together.

3   The heights of 11 students were measured in centimetres as follows:

156	151	154	155	153	154
154	153	151	60	154	

   a   Find the mode, median, mean and range for this data set.

   b   Why do you think there is a difference between the mean and the median?

   c   Which is the best average to use in this case? Why?

CONTINUED

4 A shipping company measured the mass of a sample of packing crates to the nearest kilogram in one container and produced this stem-and-leaf diagram.

Stem	Leaf
4	6
5	0 0 4
5	5 7 8 9 9
6	1 1 1 2 3
6	6 7 8 9
7	0 4

Key:
4 | 6 represents 46 kilograms

a What is the range of masses?

b What is the mode of the data?

c What is the median mass of the packing crates?

d How many crates were weighed?

e Is it accurate to say the mean mass of crates in this container is approximately 60 kg? Justify your answer.

5 Three suppliers sell specialised remote controls for access systems. A sample of 100 remote controls is taken from each supplier and the working life of each control is measured in weeks. The following table shows the mean time and range for each supplier.

Supplier	mean (weeks)	range (weeks)
A	137	16
B	145	39
C	141	16

Which supplier would you recommend to someone who wants to buy a remote control? Why?

6 A box contains 50 plastic blocks of different volumes as shown in the frequency table.

Volume (cm^3)	2	3	4	5	6	7
Frequency	4	7	9	12	10	8

a Find the mean volume of the blocks.

b What volume is most common?

c What is the median volume?

12 Averages and measures of spread

SELF ASSESSMENT

Use the answers provided to mark your work on the Review exercise.

1. How would you rate your understanding of the work? Use this rating scale to help you.

4	3	2	1
I can do the work confidently and didn't get any wrong. I can explain to someone else how to do this.	I could answer all the questions. I made a few careless mistakes, but I can see what they are.	I am starting to understand this work but there are still some areas where I need help.	I don't understand this work very well and cannot do most of it by myself.

2. Imagine you are your maths teacher. Write a few sentences commenting on this piece of work.

Chapter 13: Understanding measurement

13.1 Understanding units

KEY LEARNING STATEMENTS

- Units of measure in the metric system are metres, grams (g) and litres. Sub-divisions have prefixes such as milli- and centi-; the prefix kilo- is a multiple.

- To convert from a larger unit to a smaller unit you multiply the measurement by the correct multiple of ten.

- To convert from a smaller unit to a larger unit you divide the measurement by the correct multiple of ten.

To change to a smaller unit, multiply by conversion factor

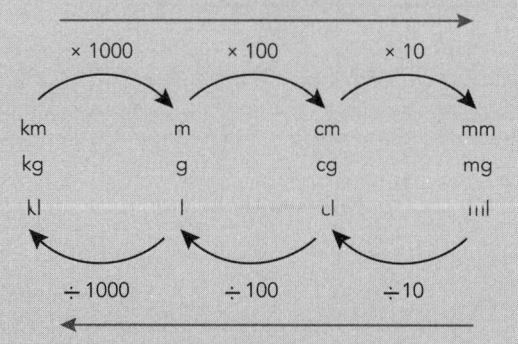

× 1000   × 100   × 10

km → m → cm → mm
kg → g → cg → mg
kl → l → cl → ml

÷ 1000   ÷ 100   ÷ 10

To change to a larger unit, divide by conversion factor.

- Area is always measured in square units. To convert areas from one unit to another you need to square the appropriate length conversion factor.

- Volume is measured in cubic units. To convert volumes from one unit to another you need to cube the appropriate length conversion factor.

KEY CONCEPT

Converting between units of measurement.

1 Use the conversion diagram in the Key learning statements box to help you draw your own diagrams to show how to convert:

   a units of area   b units of volume.

2 Convert the following length measurements to the units given.

   a 2.6 km to metres
   b 23 cm to mm
   c 8.2 metres to cm
   d 2 450 809 metres to km
   e 0.02 metres to mm
   f 15.7 cm to metres

TIP

It is useful to know these conversions:

10 mm = 1 cm
100 cm = 1 metre
1000 metres = 1 km
1000 mg = 1 g
1000 g = 1 kg
1000 kg = 1 tonne
1000 ml = 1 litre
1 cm^3 = 1 ml

3   Convert the following measurements of mass to the units given.

   a   9.08 kg to g
   b   49.34 kg to g
   c   0.5 kg to g
   d   68 g to kg
   e   15.2 g to kg
   f   2 300 000 g to tonnes

4   Identify the greater length in each of these pairs of lengths. Then calculate the difference between the two lengths. Give your answer in the most appropriate units.

   a   19 km   18 900 metres
   b   90 m   9015 cm
   c   43.3 cm   435 mm
   d   492 cm   4.29 metres
   e   635 metres   0.6 km
   f   5.8 km   580 500 cm

5   Convert the following area measurements to the units given.

   a   12 cm² to mm²
   b   9 cm² to mm²
   c   164.2 cm² to mm²
   d   0.37 km² to m²
   e   9441 m² to km²
   f   0.423 m² to mm²

6   Convert the following volume measurements to the units given.

   a   69 cm³ to mm³
   b   19 cm³ to mm³
   c   30.04 cm³ to mm³
   d   4.815 m³ to cm³
   e   103 mm³ to cm³
   f   46 900 mm³ to m³
   g   455 cm³ to litres
   h   42.25 litres to cm³

> **TIP**
> Cubic centimetres (cm³) is sometimes shortened to cc. For example, a scooter may have a 50 cc engine. This means the total volume of all cylinders in the engine is 50 cm³.

7   Naeem lives 1.2 km from school and Sadiqa lives 980 metres from school. How much closer to the school does Sadiqa live?

8   A coin has a diameter of 22 mm. If you place 50 coins in a row, how many centimetres long will the row be?

9   A square of fabric has an area of 176 400 mm². What are the lengths of the sides of the square in centimetres?

10   How many cuboid boxes, each with dimensions 50 cm × 90 cm × 120 cm, can you fit into a similar cuboid of volume 48 m³?

13.2 Time

> **KEY LEARNING STATEMENTS**
>
> - Time is not decimal. 1 h 15 mins means one hour and $\frac{15}{60}$ (or $\frac{1}{4}$) of an hour not 1.15 h.
> - You can write times using a.m. and p.m. notation or as a 24-hour time using the numbers from 0 to 24 to give the times from 12 midnight on one day (00 00) to one minutes before midnight (23 59). Even in the 24-hour clock system, time is not decimal. The time one minute after 15 59 is 16 00.

> **KEY CONCEPT**
>
> Calculating with times using the 12-hour and 24-hour clock.

> **TIP**
>
> You can express parts of an hour as a decimal. Divide the number of minutes by 60. For example,
> 12 minutes = $\frac{12}{60}$ = $\frac{1}{5}$ = 0.2 hours.
> This can make your calculations easier.

1 Five people record the time they start work, the time they finish and the length of their lunch break.

  a Work out and list how much time each person spent at work each day.

Name	Time in	Time out	Lunch	Hours worked
Dawoot	$\frac{1}{4}$ past 9	Half past five	$\frac{3}{4}$ hour	
Nadira	8.17 a.m.	5.30 p.m.	$\frac{1}{2}$ hour	
John	08 23	17 50	45 min	
Robyn	7.22 a.m.	4.30 p.m.	1 hour	
Mari	08 08	18 30	45 min	

  b Calculate each person's daily earnings to the nearest whole cent if they are paid $13.45 per hour.

2 On one particular day, the low tide in Hong Kong harbour is at 09 15. The high tide is at 15 40 on the same day. How much time passes between low tide and high tide?

3 A plane was due to land at 2.45 p.m. However, it was delayed and it landed at 15 05. How much later did the plane arrive than it was meant to?

4 How much time passes between:

  a 2.25 p.m. and 8.12 p.m. on the same day

  b 1.43 a.m. and 12.09 p.m. on the same day

  c 6.33 p.m. and 6.45 a.m. the next day

  d 1.09 a.m. and 15 39 on the same day?

5 Use this section from a bus timetable to answer the questions that follow.

North Street	09 00	09 30	10 00
South Avenue	09 18	09 48	10 18
East Place	09 35	10 05	10 35
West Lane	10 00	10 30	11 00

a What time does the earliest bus leave North Street?

b How long does the journey from North Street to West Lane take?

c The 09 30 bus is delayed by 17 minutes on its way to South Avenue. At what time will it arrive?

d Sanchez misses the 09 48 bus from South Avenue. How long will he have to wait before the next scheduled bus arrives?

e The 10 00 bus from North Street is delayed in roadworks between South Avenue and East Place for 19 minutes. Give the estimated arrival times for the rest of the timetable.

13.3 Limits of accuracy – upper and lower bounds

KEY LEARNING STATEMENTS

- All measurements we make are rounded to some degree of accuracy. The degree of accuracy (for example the nearest metre or to two decimal places) allows you to work out the highest and lowest possible value of the measurements. The highest possible value is called the upper bound and the lowest possible value is called the lower bound.

KEY CONCEPT

Upper and lower bounds to a specified degree of accuracy.

1 Each number has been rounded to the degree of accuracy shown in the brackets. Find the upper and lower bounds in each case.

a 42 (nearest whole number)

b 13 325 (nearest whole number)

c 400 (1 sf)

d 12.24 (2 dp)

e 11.49 (2 dp)

f 2.5 (nearest tenth)

g 390 (nearest ten)

h 1.132 (4 sf)

TIP

dp means decimal places

sf means significant figures

2 A building is 72 metres tall measured to the nearest metre.

a What are the upper and lower bounds of the building's height?

b Is 72.499 999 999 999 999 999 metres a possible height for the building? Explain why or why not.

13.4 Conversion graphs

> **KEY LEARNING STATEMENTS**
>
> - Conversion graphs allow you to convert from one unit of measure to another by providing the values of both units on different axes. To find one value when the other is given, you must find the point on the graph for the given value and then read off the corresponding value on the other axis.

KEY CONCEPT

- Interpreting and using conversion graphs.
- Graphs in practical situations.

TIP

Make sure you read the labels on the axis so that you are reading off the correct values.

1   An Australian tourist is going on holiday to the island of Bali in Indonesia. This conversion graph shows the value of rupiah (the currency of Indonesia) against the Australian dollar.

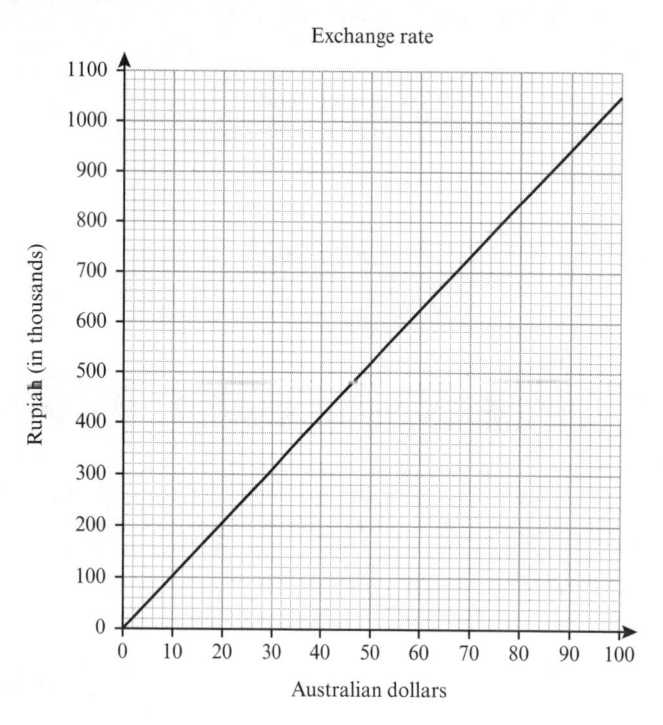

Exchange rate

(Graph: vertical axis "Rupiah (in thousands)" from 0 to 1100; horizontal axis "Australian dollars" from 0 to 100; straight line from origin through approximately (100, 1050).)

a   What is the scale on the vertical axis?

b   Convert these amounts to rupiah.

   i   Aus $50   ii   Aus $100   iii   Aus $500

c   The tourist finds a hotel for 400 000 rupiah a night.

   i   What is this amount in Australian dollars?

   ii   How much will it cost in Australian dollars for an eight-night stay?

2   Study the conversion graph and answer the questions.

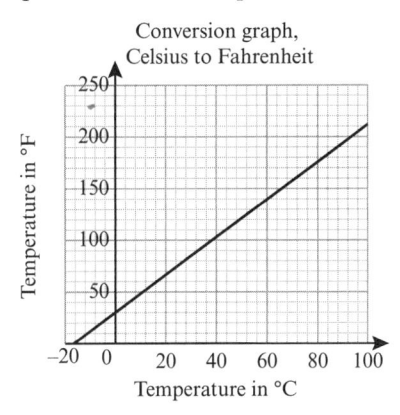

Conversion graph, Celsius to Fahrenheit

a   What is shown on the graph?

b   What is the temperature in Fahrenheit when it is:

   i   0 °C   ii   10 °C   iii   100 °C?

c   Sarah finds a recipe for chocolate brownies that says she needs to cook the mixture at 210 °C for one hour. After an hour she finds that it has hardly cooked at all. What could the problem be?

d   Jess is American. When she calls her friend James in England she says, 'It's really cold here, must be about 50 degrees out.' What temperature scale is she using? How do you know this?

TIP

The USA mainly still uses the Fahrenheit scale for temperature. Appliances, such as ovens, may have temperatures in Fahrenheit on them, particularly if they are an American brand.

3   This graph shows the conversion factor for pounds (imperial measurement of mass) and kilograms.

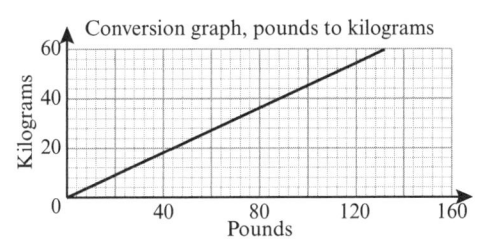

Conversion graph, pounds to kilograms

a   A doctor tells a patient they need to lose 20 pounds. How much is this in kilograms?

b   A student weighs 98 pounds. How much is this in kilograms?

c   Which is the greater mass in each of these cases:

   i   30 pounds or 20 kg

   ii   35 kg or 70 pounds

   iii 60 kg or 145 pounds?

13.5 Exchanging currencies

KEY LEARNING STATEMENTS

- When you change money from one currency to another you do so at a given rate of exchange. Changing to another currency is called buying foreign currency.
- Exchange rates can be worked out using conversion graphs (as in Exercise 13.4), but more often, they are worked out by doing calculations.
- Doing calculations with money is just like doing calculations with decimals but you need to remember to include the currency symbols in your answers.

KEY CONCEPT

Calculating with money and converting between different currencies.

Use the exchange rate table below for these questions.

Currency exchange rates

Currency	US $	Euro €	UK £	Indian rupee	Aus $	Can $	SA rand	NZ $	Yen ¥
1 US $	1.00	0.90	0.75	76.00	1.37	1.26	15.28	1.48	115.76
inverse	1.00	1.11	1.34	0.01	0.73	0.79	0.07	0.68	0.01
1 Euro	1.11	1.00	0.83	84.25	1.52	1.40	16.93	1.64	128.32
inverse	0.90	1.00	1.21	0.01	0.66	0.71	0.06	0.61	0.01
1 UK £	1.34	1.21	1.00	101.58	1.83	1.69	20.42	1.97	154.71
inverse	0.75	0.83	1.00	0.01	0.55	0.59	0.05	0.51	0.01

TIP

The inverse rows show the exchange rate of one unit of the currency in the column to the currency above the word inverse. For example, using the inverse row below US$ and the euro column, €1 will buy $1.11.

1 What is the exchange rate for:

    a US$ to yen   **b** UK£ to NZ$

    c Euro to Indian rupee   **d** Canadian dollar to euro

    e Yen to British pound   **f** South African rand to US$?

2 How many Indian rupees will you get for:

    a US$50   **b** 600 euros   **c** £95?

3 How many yen will you need to buy:

    a US$120   **b** 500 euros   **c** £1200?

TIP

Currency rates change all the time. You need to read tables carefully and use the rates that are given.

SELF ASSESSMENT

A learning log can help you identify what revision you need to do to improve your understanding and performance.

Learning log for:	*Converting between currencies*
I learned …	
I still need to learn more about ..	
My next steps will be …	
I need some help with …	
In general, I feel _____ about this section of work.	

Copy the headings and complete the learning for the work on converting between one currency and another.

REVIEW EXERCISE

1. Convert the following measurements to the units given.

   a 2.7 km to metres
   b 69 cm to mm
   c 6 tonnes to kilograms
   d 23.5 grams to kilograms
   e 263 grams to milligrams
   f 29.25 litres to millilitres
   g 240 cm³ to litres
   h 10 cm² to mm²
   i 6428 m² to km²
   j 7.9 m³ to cm³
   k 0.029 km³ to m³
   l 168 mm³ to cm³

2. The average time taken to walk around a track is one minute and 35 seconds. How long will it take you to walk around the track 15 times at this rate?

3. A journey took 3 h 40 min and 10 s to complete. Of this, 1 h 20 min and 15 s was spent having lunch or stops for other reasons. The rest of the time was spent travelling. How much time was actually spent travelling?

4. Tayo's height is 1.62 metres, correct to the nearest centimetre. Calculate the least possible and greatest possible height that Tayo could be.

5. The number of people who attended a meeting was given as 50, correct to the nearest 10.

   a Is it possible that 44 people attended? Explain why or why not.
   b Is it possible that 54 people attended? Explain why or why not.

CONTINUED

6 Study the graph and answer the questions.

  a What does the graph show?

  b Convert to litres:

   i 10 gallons   **ii** 25 gallons.

  c Convert to gallons:

   i 15 litres   **ii** 120 litres.

  d Naresh says he gets 30 mpg in the city and 42 mpg on the highway in his car.

   i Convert each rate to kilometre per gallon.

   ii Given that one gallon is equivalent to 4.546 litres, convert both rates to kilometres per litre.

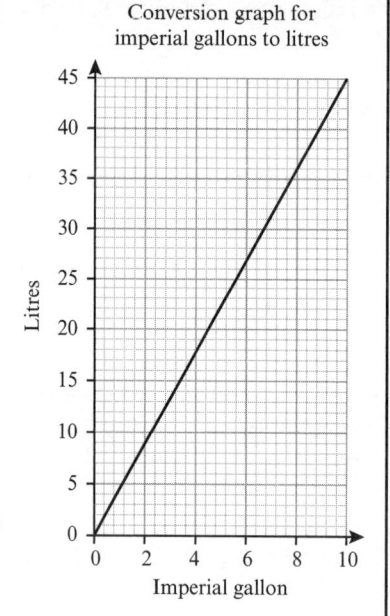

TIP

The US gallon (customary measure) is different from the imperial gallon with a conversion factor of 1 US gallon to 3.785 litres.

TIP

1 mile = 1.61 km

Use the exchange rate table (repeated from Exercise 13.5) to answer the following questions.

Currency exchange rates

Currency	US $	Euro €	UK £	Indian rupee	Aus $	Can $	SA rand	NZ $	Yen ¥
1 US $	1.00	0.90	0.75	76.00	1.37	1.26	15.28	1.48	115.76
inverse	1.00	1.11	1.34	0.01	0.73	0.79	0.07	0.68	0.01
1 Euro	1.11	1.00	0.83	84.25	1.52	1.40	16.93	1.64	128.32
inverse	0.90	1.00	1.21	0.01	0.66	0.71	0.06	0.61	0.01
1 UK £	1.34	1.21	1.00	101.58	1.83	1.69	20.42	1.97	154.71
inverse	0.75	0.83	1.00	0.01	0.55	0.59	0.05	0.51	0.01

7 A South African tourist is going on holiday to Italy and wants to exchange R10 000 for euros. How many euros will they get?

8 An American travelling to India for business needs to exchange $2000 for rupees.

  a What is the exchange rate?

  b How many rupees will she get at this rate?

  c At the end of the trip she has 12 450 rupees left over. What will she get if she changes these back to dollars at the given rate?

9 A British family is going to Spain on holiday. The cost of the holiday is 4875 euros. What is this amount in UK pounds?

Chapter 14: Further solving of equations and inequalities

14.1 Simultaneous linear equations

> **KEY LEARNING STATEMENTS**
>
> - Simultaneous equations give information about two unknowns (x and y). You have to solve them together (simultaneously) to find values for x and y which make both equations true at the same time.
>
> - You can represent the equations linking x and y as straight line graphs. The coordinates (x, y) of the point where the lines intersect give the solution to the equations.
>
> - You can also solve simultaneous equations algebraically using either substitution or elimination. You may need to rearrange the equations before you do this.

> **KEY CONCEPT**
>
> Solving simultaneous equations in two unknowns.

1 Each diagram shows the graphs of two linear equations. Use the graphs to find the simultaneous solution to each pair of equations.

a Lines: $x + y = 5$ and $2x - y = 4$

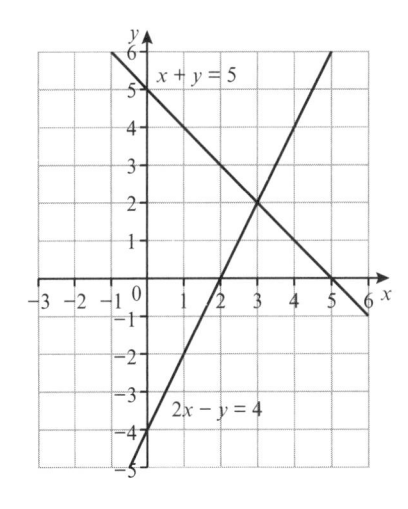

b Lines: $y = 2x$ and $y = -x + 3$

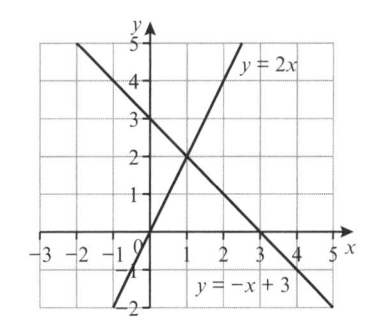

c

$2x + y = 5$
$x - 3y = 6$

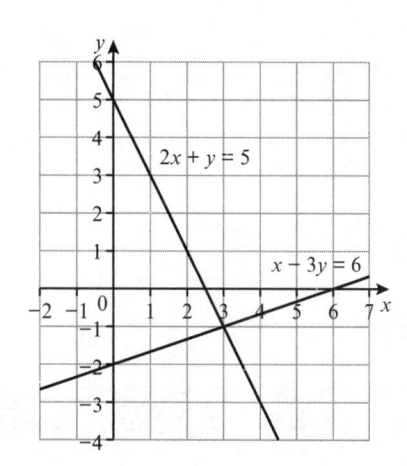

d

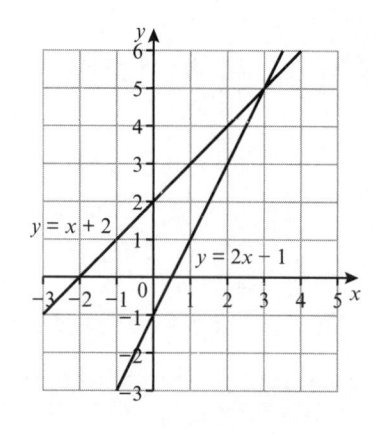

$y = x + 2$
$y = 2x - 1$

2 For each pair of equations, draw suitable graphs and use them to find the simultaneous solution to the equations.

    a $y = 3x - 5$ **b** $x + y = 2$ **c** $2x + y = 12$ **d** $3x - 2y = 5$

         $y = -2x + 5$ $x - 3y = 6$ $y = x - 3$ $4x = 18 - 3y$

3 Five graphs have been drawn on the same set of axes.

    a Determine the equation of each line A to E.

    b Find the simultaneous solution for each of the following pairs of equations:

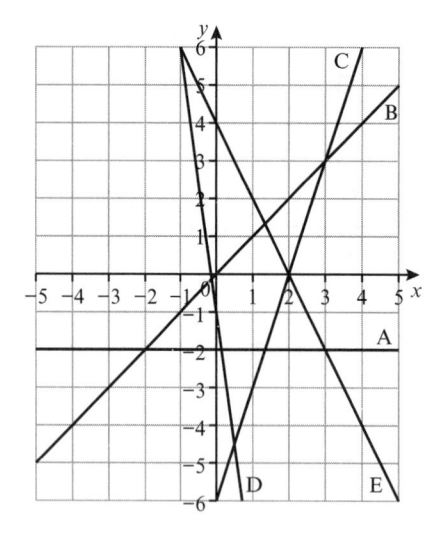

        i   A and B   **ii**   B and C   **iii**   A and E

        iv   D and E   **v**   C and E.

4 Solve for x and y using the substitution method.

    a $y = 4 - x$ **b** $y = 7 - x$ **c** $x + y = 4$

         $y = 2x + 1$ $x + 3y = 27$ $x + 3y = 6$

    d $2x + 3y = 8$ **e** $y = 4x - 2$ **f** $3x + 5y = 21$

         $x + 4y = 9$ $3x - 5y = 27$ $2x + y = 7$

> **TIP**
>
> For the substitution method, make x or y the subject of one of the equations and substitute what x or y is equal to into the other equation.

5 Use the elimination method to solve each pair of equations for x and y.

a $y = 2x - 5$
$y = -2x + 3$

b $x + y = 5$
$x - y = 3$

c $3x + 2y = 6$
$-3x + y = 0$

d $x - 5y = 19$
$3x + 5y = -3$

e $2x + 5y = 9$
$x + 3y = 5$

f $7x + 2y = 1$
$4x + 3y = 8$

6 Solve the simultaneous equations algebraically. Use the method you think is most appropriate for each pair.

a $y = x - 3$
$y = -2x$

b $x + y = 3$
$x - y = 1$

c $x + y = 4$
$x + 3y = 6$

d $2x + y = 12$
$x - y = 3$

e $3x - 2y = 29$
$y = 24 - 4x$

f $6x - 7y = 16$
$-2y + 3x = 5$

g $2x + 3y = 12$
$-4y = 1 - 3x$

h $2x + 4y = 18$
$-y = 3 - 2x$

i $4x + 3y = 5$
$5x + 6y = 4$

j $7x + 2y = 37$
$6x = 27 + 3y$

7 Two numbers, x and y, have a sum of 120. When y is subtracted from $3x$, the difference is 160. Determine the value of x and y.

8 Three packs of markers and two notebooks have a combined mass of 610 g. Five packs of markers and four notebooks have a combined mass of 1070 g. Determine the mass of each notebook and the mass of each marker.

9 A large yoga studio offers a number of different classes. They charge $8 for a class and they pay the instructors $15 each for teaching. During one session there are 23 people either taking a class or teaching a class and the studio makes a profit of $92 after the instructors have been paid. How many people took a class during this session?

> **TIP**
>
> For the elimination method, write the equations one above the other. If necessary, multiply one or both of the equations by a constant to make the x- or y-coefficients equal and then add or subtract the equations to eliminate one of the variables.

14.2 Linear inequalities

> **KEY LEARNING STATEMENTS**
>
> - An inequality is the relationship between two unequal expressions.
> - The symbols $<, >, \leq, \geq$ and $\neq$ all indicate inequalities.
> - Inequality symbols can be used to show a set of values. $3 < x \leq 10$ is the set of values that are greater than three and also smaller than or equal to ten.
> - You can show the solutions to inequalities on a number line. Use a coloured in circle to show a value is included and an open circle to show a value is not included.

> **KEY CONCEPT**
>
> Representing and interpreting inequalities.

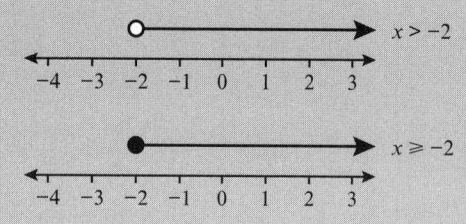

1 Write an inequality for each statement.

   a x is greater than nine.

   b y is less than -5.

   c x is equal to one or less.

   d y is greater than -2 but less than six.

   e x is less than -4 but greater than or equal to -10.

   f Three more than y is greater or equal to 4 less than x.

2 Write down two integer values that will satisfy each of these inequalities.

   a $x \leq 9$ **b** $9 < y$ **c** $3 > x$

   d $2 < x < 7$ **e** $x \geq 300$ **f** $x \leq 0$

   g $-4 < x \leq 0$ **h** $-1.4 < x < 3.8$ **i** $x > 5$ or $x \leq 3$

3 Write the inequality represented by each number line.

   a, **b**, 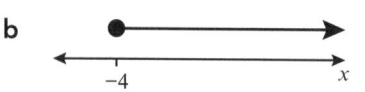 **c**, 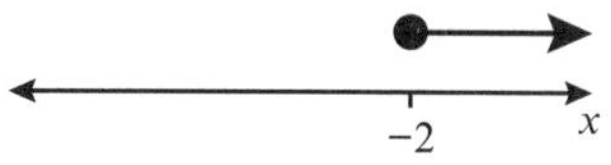 **d**, 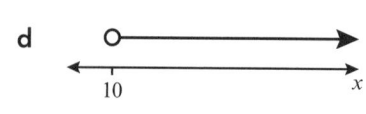 **e**, 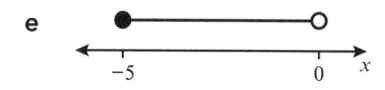 **f**

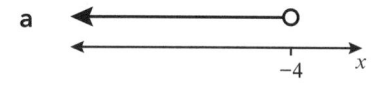

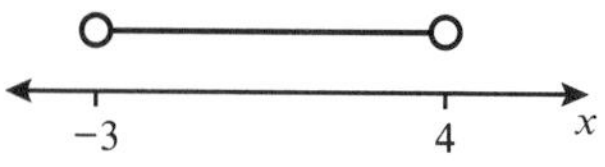

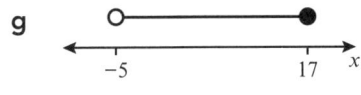

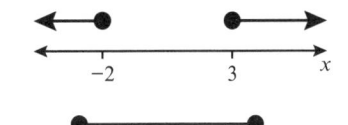

14 Further solving of equations and inequalities

g ○———● between −5 and 17, x

h ←● −2, ●→ 3, x

i ←○ −2, ●→ $2\frac{1}{2}$, x

j ●———● 2.7 to 6.3, x

4 Draw a number line to represent each inequality.

a $x < 6$
b $x > -3$
c $y \geq -5$
d $y \leq -3$
e $1.2 < x < 4.8$
f $-3.5 < y \leq 2.8$
g $-8 \leq x \leq -3$
h $x < 15$ or $x \leq 17$
i $x \leq 3$ or $x \geq 9$

TIP

You only need to show the values representing the inequalities, you do not need to show all the divisions on the number line.

SELF ASSESSMENT

How well do you understand linear inequalities and how to show them on a number line?

1 Mark your own answer to questions 1 to 4.

2 Use this scale to give yourself a rating of 1, 2, 3 or 4.

4	3	2	1
I know what we are trying to do. I understand how to read the symbols and work out the possible values. I can show the inequality on a number line and use the correct markings. I can clearly explain this work to someone else.	I mostly know what we are trying to do. I understand most of the work on inequalities. I can draw number lines to show most of the inequalities but I find some questions confusing.	I think I need more practice. I get confused and am not sure what some of the inequalities represent. I cannot do many of the questions.	I really don't understand this at all. I need help with all aspects of this work.

3 If you rated yourself 3, 2 or 1, write down the next step you will take to improve your rating.

REVIEW EXERCISE

1 Draw accurate graphs and use them to solve each pair of simultaneous equations given below. Check your answers by substituting the coordinates in the original.

    a $y = x + 3$
          $x + 4y = 22$

    b $y = x - 2$
          $x + 4y = 12$

    c $y = x + 1$
          $x + y = 2$

    d $y = \frac{1}{2}x - 1$
          $x + 2y = 0$

2 Solve for x and y if $7x + y = 8$ and $-x + y = -8$.

3 Find the simultaneous solution for $3x + 2y - 1 = 0$ and $-2x + y + 3 = 0$.

4 The sum of x and y is 38 and the difference between x and y is 24. What are the values of x and y?

5 Five melons and three peaches cost $14.05, three melons and two peaches cost $8.50. What is the price of each type of fruit?

6 The diagram shows two intersecting lines.

    a   Write the equation of each line.

    b   Read the simultaneous solution to the equations from the graph.

    c   Show algebraically that your solution is correct.

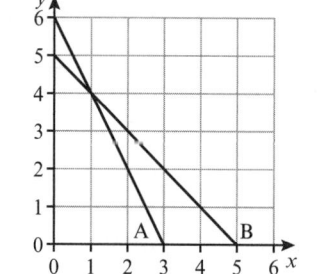

7 Find the perimeter of this rectangle.

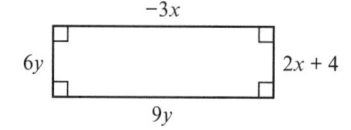

(Rectangle with sides labelled $-3x$, $2x + 4$, $9y$, $6y$.)

8 Write an inequality and draw a number line to represent each of the following situations. Use the variables that are given.

    a   The base (b) of a triangle is less than 15 mm long.

    b   The length (l) of a field is no longer than 11 metres.

    c   An event was attended by up to 500 people (p).

    d   There were between 1200 and 1500 people (p) at a soccer match.

    e   I spend $180 or more on utilities (u) every month but never more than $250.

    f   In a test, I scored at least 60 marks (m) but less than 75 marks.

Chapter 15: Scale drawings, bearings and trigonometry

15.1 Scale drawings

> **KEY LEARNING STATEMENTS**
> - Scale drawings show the shape and position of objects and to allow you to measure distances accurately. Maps and building plans are scale drawings.
> - The scale of a drawing is the ratio of length on the drawing to actual length. If the scale is 1 : 10, the drawing has lengths that are $\frac{1}{10}$ of the real thing. In this example, the scale factor is 10.
> - You can work out the lengths you need for your scale drawing using the rule: length on drawing = real length ÷ scale factor.
> - You can calculate real distances from the lengths on the scale drawing using the rule: real distance = length on drawing × scale factor.

> **KEY CONCEPT**
> Drawing and interpreting scale diagrams.

1 Use a scale of 1 : 5 to make a scale drawing of this pattern.

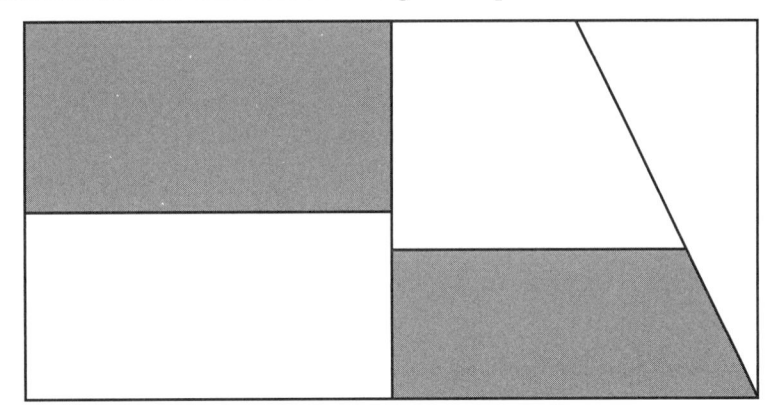

2 Make a scale drawing of this shape using a scale of 3 : 1.

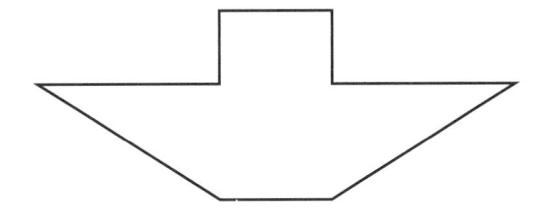

117

3 A surveyor wants to draw a scale diagram of a rectangular field that is 20 metres wide and 50 metres long.

a How many millimetres long will the field be on the diagram if the surveyor uses a scale of:

i 1 cm to 5 metres

ii 4 mm to 1 metre

iii 1 : 200

iv 1 : 400?

b Which of these scales is best if the surveyor draws the plan on an A4 sheet of paper?

4 A plan has a scale of 1 : 400. Calculate the real distance represented by a plan length of:

a 4 cm **b** 25 mm **c** 3.1 cm **d** $\frac{1}{2}$ cm.

5 A locust is drawn using a scale of 1 : 0.5. A leg on the diagram is 2.6 cm long. How long is the real leg?

6 Find the real length of the *Paramecium* (a single-celled organism) using the scale shown.

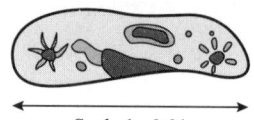

Scale 1 : 0.01

15.2 Bearings

KEY LEARNING STATEMENTS

- You can give compass directions using north, south, west and east.

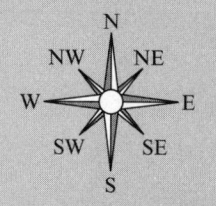

- For more specific directions you can use degrees or bearings.

- Bearings are measured from north (000°) in a clockwise direction to 360° (which is back at north).

- The diagram shows the bearings from A to B and from B to A. Note that you measure each bearing from north of the starting point.

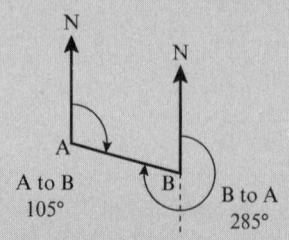

A to B 105°

B to A 285°

- You write bearings using three figures, so an angle of 58° is a bearing of 058°.

KEY CONCEPT

Using and interpreting three-figure bearings.

15 Scale drawings, bearings and trigonometry

1 The scale diagram shows the location of four places. Make a copy of this diagram using 5 mm squared paper.

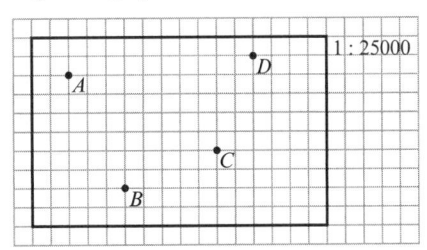

> **TIP**
> You use a protractor to measure bearings. Put the 0° line of the protractor in line with north and then measure the angle.

a Which place is south-west of *D*?

b What is the bearing from

    i *A* to *C*?   **ii** *B* to *A*?

c Place *E* is 1.2 km from *A* on a bearing of 135°. What is the bearing from *E* to *D*?

2 On a map, the village of Parakou is 5 cm north and 3 cm west of the village of Ede. What is the bearing of Parakou from Ede?

3 A surfskier paddled in a straight line to reach a buoy 160 metres north and 120 metres east of its starting point. By drawing a scale diagram, calculate:

a how far the surfskier paddled

b on what bearing the surfskier paddled.

4 Two people start walking from the same point. The first person walks due east for 3.5 km. The second person walks on a bearing of 144° until they are due south of the first person. Use a scale diagram to work out how far the second person walked. Give your answer to the nearest whole kilometre.

15.3 Understanding the tangent, cosine and sine ratios

KEY LEARNING STATEMENTS

- The sides of a right-angled triangle are given specific names in relation to a known or given angle (θ).

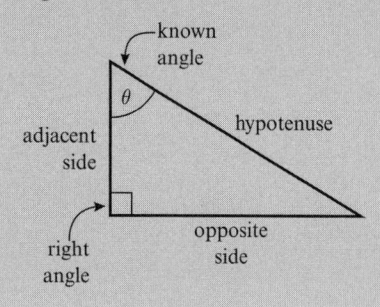

- Trigonometric ratios (tangent, cosine and sine) describe the relationship between a pair of sides and the given angle. You can use these ratios to find the lengths of unknown sides or the sizes of unknown angles in right-angled triangles.

- For angle θ, the trigonometric ratios are:

$$\sin\theta = \frac{\text{opp}(\theta)}{\text{hyp}}, \quad \cos\theta = \frac{\text{adj}(\theta)}{\text{hyp}} \quad \text{and} \quad \tan\theta = \frac{\text{opp}(\theta)}{\text{adj}(\theta)}$$

- To find the size of unknown angles, use the inverse functions ($\sin^{-1}$, $\cos^{-1}$ and $\tan^{-1}$).

- To find the length of an unknown side choose the appropriate ratio and rearrange it to make and solve an equation. For example, if you know the angle and the opposite side and you want to find the hypotenuse, rearrange $\sin\theta = \frac{\text{opp}(\theta)}{\text{hyp}}$ to give $\text{hyp} = \frac{\text{opp}(\theta)}{\sin\theta}$, substitute values for opp (θ) and θ, and solve for the hypotenuse.

KEY CONCEPT

Using the sine, cosine and tangent ratios to solve right-angled triangles.

1 Copy and complete this table for the given triangles.

Triangle	Hypotenuse	Opposite θ	Adjacent θ
ABC	AB		
DEF			
XYZ			

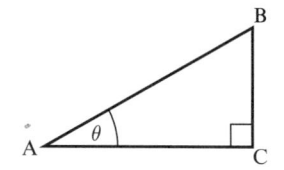

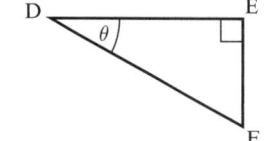

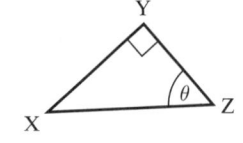

TIP

Make sure your calculator is in degree mode when you are using the trigonometric function keys.

TIP

The Greek letter theta (θ) is often used to represent the given angle.

15 Scale drawings, bearings and trigonometry

2 For each triangle, find:

   i $\sin\theta$ **ii** $\cos\theta$ **iii** $\tan\theta$

   Give your answer as a fraction.

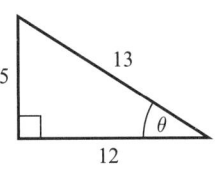

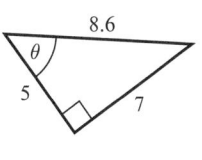

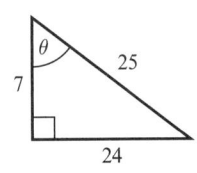

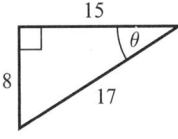

   a (triangle with sides 3, 4, 5 and θ at top)

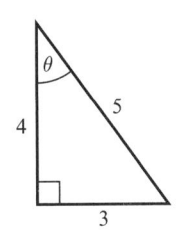

   b (triangle with sides 5, 12, 13 and θ at bottom right)

   c (triangle with sides 5, 7, 8.6 and θ at top left)

   d (triangle with sides 7, 24, 25 and θ at top left)

   e (triangle with sides 8, 15, 17 and θ at top)

   f (triangle with sides 9, 12, 15 and θ at left)

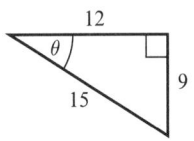

3 Use your calculator to find, correct to three decimal places, the value of the following.

   a $\sin 48°$ **b** $\cos 12°$ **c** $\tan 69°$ **d** $\sin 24.6°$

   e $\cos 33°$ **f** $\tan 40°$ **g** $\tan 10.5°$ **h** $\cos 6.87°$

4 For each triangle, find the length of x, correct to two decimal places.

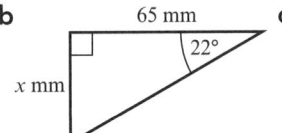

   a (right triangle, hypotenuse 8 cm, angle 46°, opposite side x cm)

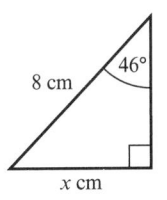

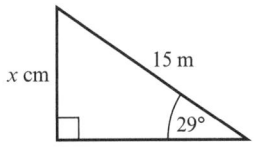

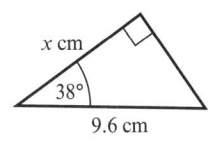

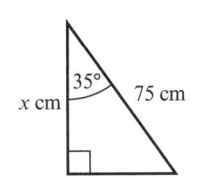

   b (right triangle, 65 mm top, angle 22°, x mm left side)

   c (right triangle, hypotenuse 15 m, angle 29°, x cm)

   d (right triangle, 9.6 cm, angle 38°, x cm)

   e (right triangle, hypotenuse 75 cm, angle 35°, x cm)

   f (right triangle, 8 m, angle 21°, x m)

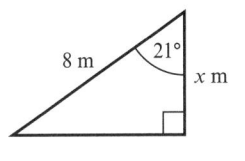

5 Given that θ is acute, find the size of θ to the nearest degree if:

   a $\tan\theta = 0.635$ **b** $\sin\theta = 0.2135$ **c** $\cos\theta = 0.7231$

   d $\sin\theta = 0.632$ **e** $\cos\theta = 0.2954$ **f** $\tan\theta = 1.2$

6 Find the size of the marked angle. Give your answer correct to one decimal place.

a, b, 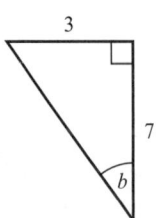 c, 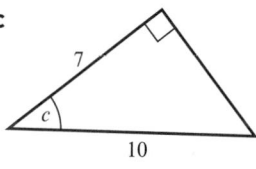 d, 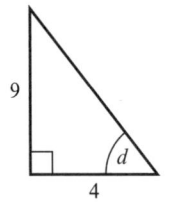 e, 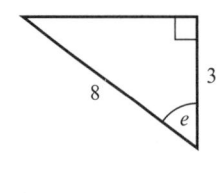 f (triangle diagrams)

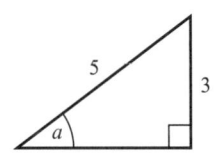

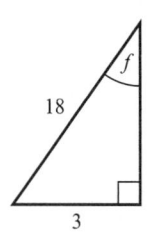

> **TIP**
>
> The memory aid, SOHCAHTOA, or the triangle diagrams may help you remember the trigonometric relationships.

15.4 Solving problems using trigonometry

KEY LEARNING STATEMENTS

- You can solve problems involving right-angled triangles using trigonometry and/or Pythagoras' theorem.
- If you are not given a diagram, draw your own and label all the given sides and angles. Mark any angles or sides that you need to find.
- Identify the right-angled triangles you can use to find angles or sides and decide whether to use Pythagoras' theorem or trigonometric ratios to solve the problem.

KEY CONCEPT

Solving problems in two dimensions using Pythagoras' theorem and trigonometry.

1 A pole 12.5 metres high is secured by two wire ropes which make an angle of 56° with the ground.

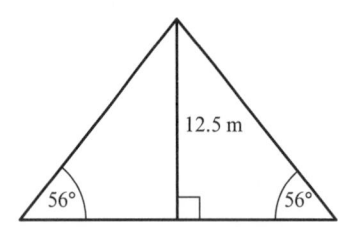

a What is the length of each wire, correct to the nearest centimetre?

b A pole twice as high is secured in the same way. If the angle between the ground and the stay wire remains 56°, how long will each wire be?

2   A diver is secured to the dive boat by a safety line which is let out as the diver moves. At a depth of 3.2 metres, the safety line makes an angle of 52° with the surface of the water and there is 2 metres of safety line above the surface. What is the total length of the safety line let out so far? Give your answer in metres, correct the nearest centimetre.

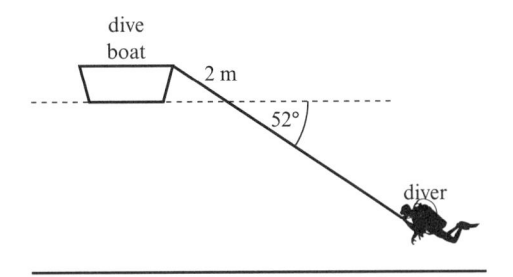

3   The diagonal of a square is 23.5 cm long. Calculate the length of a side, correct to two decimal places.

4   A small plane flies 64 km on a bearing of 215°. How far south is the plane from its original position at this point? Give your answer correct to two decimal places.

5   An environmentalist launches a specialised camera drone on a bearing of 120° from point X. The drone moves at a speed of 650 metres per hour. After three hours, how far is the drone:

   a   east of point X
   b   south of point X?

REFLECTION

Some students find working with trigonometric ratios confusing.

Do you find anything confusing in this section? If so, what did you do?

How confident are you using your calculator to find trigonometric ratios and their inverses?

Which concepts in this section would you like to understand better? How will you achieve that?

REVIEW EXERCISE

1   Draw lines (in cm) to show how long each of the following lengths would be on a scale drawing with a scale of 1 : 250.

   a   250 cm
   b   500 cm
   c   850 cm
   d   3.5 metres
   e   9000 mm
   f   0.0045 km

CONTINUED

2 The bearings of five aeroplanes, all 200 km from a control tower, are given below. Copy the diagram and accurately indicate the position of each plane on the diagram.

    i   065°   **ii**   093°   **iii**   172°

    iv   268°   **v**   308°

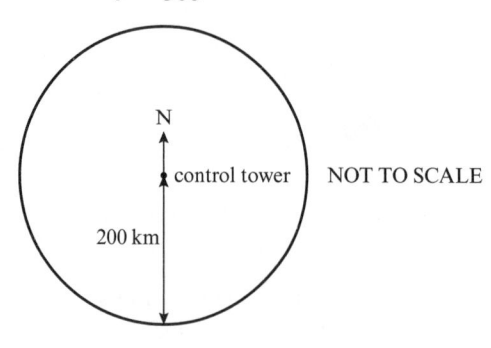

3 If the bearing of X from Y is 050° and the bearing of Y from W is 030°, calculate the size of:

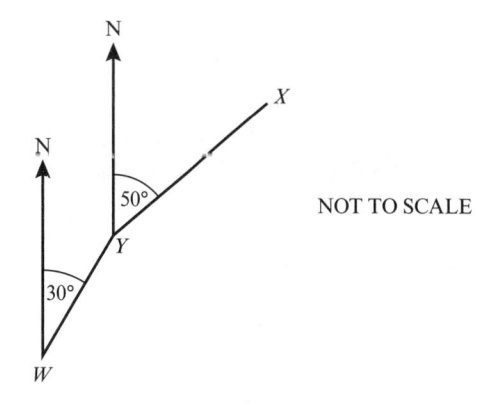

    a   angle WYN **b**   angle WYX

4 For the given triangle XYZ, work out the value of:

    a $\sin X$ **b** $\cos X$ **c** $\tan X$

    d $\sin Y$ **e** $\cos Y$ **f** $\tan Y$.

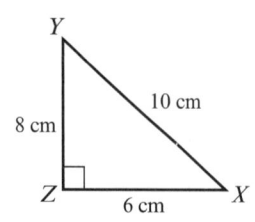

CONTINUED

5 For each of the following choose the correct trigonometric ratio and work out the value of x correct to two decimal places.

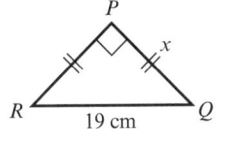

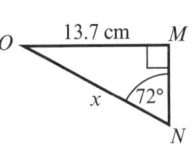

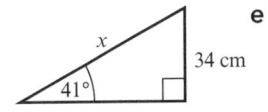

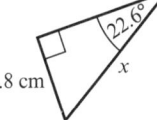

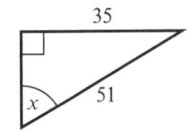

a Triangle ABC right-angled at B, angle $A = 30°$, $AB = 25$ cm, $CB = x$.

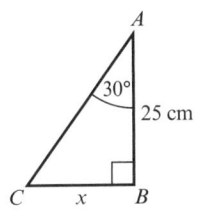

b Triangle PQR right-angled at P, with $PR = PQ$ (marked equal), $RQ = 19$ cm, x is from P to Q (side PQ).

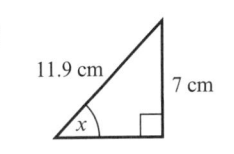

c Triangle OMN right-angled at M, $OM = 13.7$ cm, angle $N = 72°$, $ON = x$.

d Right triangle with angle $41°$, hypotenuse x (opposite the right angle side labelled), side 34 cm.

e Right triangle with angle $22.6°$, side 4.8 cm, x opposite.

f Right triangle with hypotenuse 35, side 51, x opposite right angle adjacent.

g Right triangle, hypotenuse 11.9 cm, side 7 cm, angle x.

h Right triangle, top side 7 cm, left side 11.5 cm, angle x.

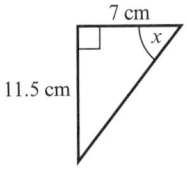

6 A hiking path slopes upwards at an angle of 18°.

If a hiker walks 600 metres up the path, how high are they above the starting point?

7 A 9 metre high vertical mobile phone mast on a level concrete slab is supported by two wires 10 metres long. Each wire is attached to the top of the pole and to the slab.

Calculate:

a The angle between the wire and the slab.

b The distance from the bottom of the mast to the point where the wires are attached to the slab.

8 The diagram shows two ladders placed in an alley, both reaching up to window ledges on opposite sides: one is twice as long as the other and reaches height H, while the shorter one reaches a height h. The length of the longer ladder is 7.5 metres. If the angles of inclination of the ladders are as shown, what is the difference in height between the window ledges?

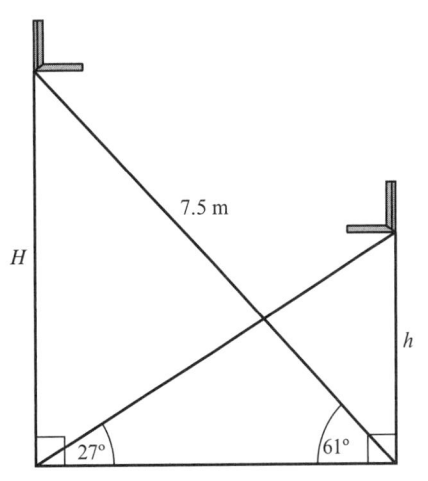

Chapter 16: Scatter diagrams and correlation

16.1 Introduction to bivariate data

KEY LEARNING STATEMENTS

- When you collect two sets of data in pairs, it is called bivariate data. For example, you could collect height and mass data for various students.

- You can plot bivariate data on a scatter diagram in order to look for correlation – a relationship between the data. For example, if you wanted to know whether taller students weighed more than smaller students, you could plot the two sets of data (height and mass) on a scatter diagram.

- Describe correlation as positive or negative, and strong or weak. When the points follow no real pattern, there is zero correlation.

- You can draw a line of best fit on a scatter diagram to describe the correlation. This line should follow the direction of the points on the graph and there should be more or less the same number of points on each side of the line.

- You can use a line of best fit to make predictions within the range of the data shown. It is not statistically accurate to predict beyond the values plotted.

KEY CONCEPTS

- Drawing and interpreting scatter diagrams.
- Types of correlation shown on scatter diagrams.
- Drawing and interpreting a line of best fit.

1 Describe the correlation shown on each scatter diagram.

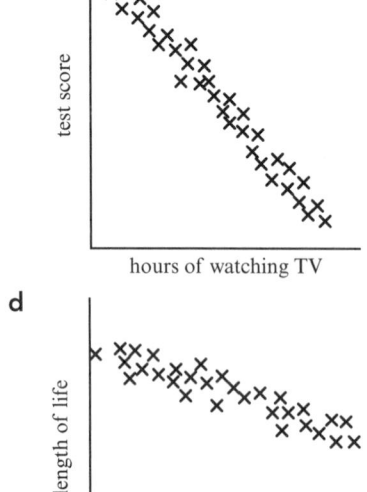

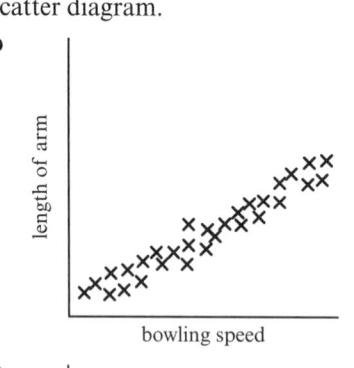

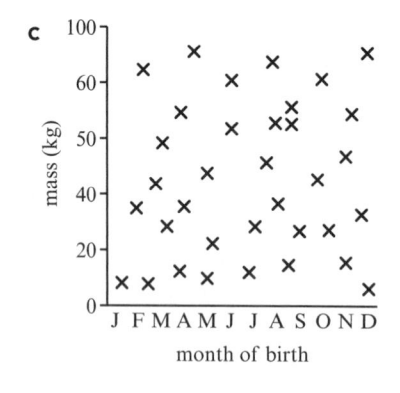

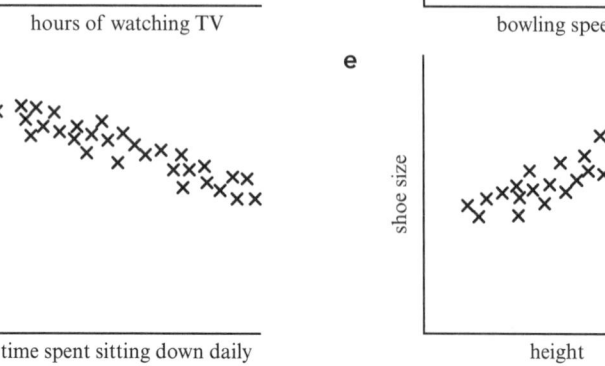

a. test score vs hours of watching TV
b. length of arm vs bowling speed
c. mass (kg) vs month of birth
d. length of life vs time spent sitting down daily
e. shoe size vs height

2 Sookie collected data from 15 students in her school athletics team. She wanted to see if there was a correlation between the height of the students and the distance they could jump in the long jump event. She drew a scatter diagram to show the data.

Student heights compared to distance jumped

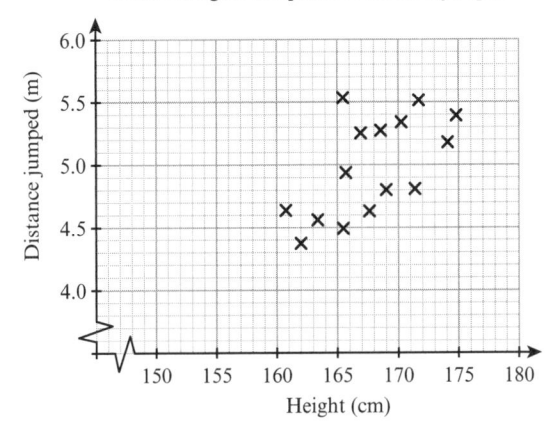

a Copy the diagram and draw the line of best fit on to it.

b Use your line of best fit to estimate how far a student 165 cm tall can jump.

c For the age group of Sookie's school team, the record for long jump is 6.07 metres.

   i   How tall would you predict a student to be who could equal the record jump?

   ii Explain whether you think this is a reliable prediction.

d Describe the correlation shown on the graph.

e What does the correlation indicate about the relationship between height and distance jumped in the long jump event?

TIP

When you plot points on a graph, it is important to mark them clearly. Use a sharp pencil and indicate each point plotted using a small x mark.

3 A group of students tasted 11 different brands of fruit juice and rated them one to ten based on how good they tasted. The price per litre of each brand of juice and the rating that the students gave is given in the table.

Brand	A	B	C	D	E	F	G	H	I	J	K
Price per litre ($)	1.50	1.70	1.50	2.50	2.00	1.20	1.85	1.00	2.80	2.20	2.95
Taste rating	4	6	9	10	6	6	8	4	9	7	9

a Plot a scatter diagram to show this data.

b Add a line of best fit and describe the correlation.

c Is it fair to say that the more expensive juices tasted better? Give a reason for your answer.

> CAMBRIDGE IGCSE™ MATHEMATICS: CORE PRACTICE BOOK

SELF ASSESSMENT

Answer the questions. Give your answer using one of these:

      Yes   Yes, but ...   No, but ...   No

1. Can you draw a scatter diagram?
2. Are you able to read scatter diagrams and describe the relationship between the variables shown?
3. Do you understand what negative, positive and zero correlation mean?
4. Can you draw a line of best fit?
5. Are you able to make predictions using the data on a scatter diagram?

Write short action points to improve in the areas where you didn't answer yes.

REVIEW EXERCISE

1. Study the scatter diagram and answer the questions.

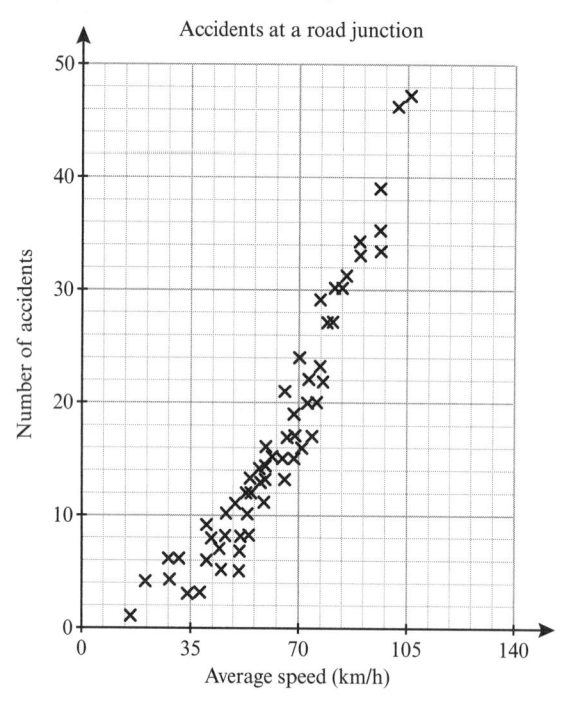

   Accidents at a road junction

   a. What does this diagram show?

   b. What is the independent variable?

CONTINUED

    c   Use a ruler to model the line of best fit. Use the line to predict:

        i   the number accidents at the junction when the average speed of vehicles is 100 km/h

        ii   the average speed of vehicles for which there are fewer than 10 accidents.

    d   Describe the correlation.

    e   What does your answer to part (**d**) tell you about the relationship between speed and the number of accidents at a junction?

2   A brand new car costs $15 000. Mr. Smit wants to find out what price second-hand cars of the same model are sold for. He drew this scatter diagram to show the relationship between the price and the age of cars in a second-hand car dealership.

Comparison of car age with re-sale price

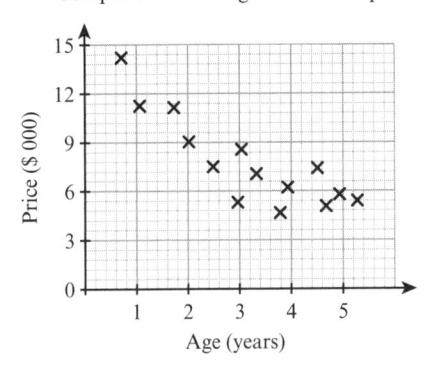

    a   Describe the trend shown on this graph.

    b   In which two-year period does the price of the car fall in value by the largest amount?

    c   Describe what happens to the price when the car is three to five years old.

    d   How old would you expect a second-hand car to be if it was advertised for sale at $7500?

    e   What price range would you expect a three-year-old car to fall into? How reliable do you think your answer is?

Chapter 17: Managing money

17.1 Earning money

KEY LEARNING STATEMENTS

- People who are formally employed may be paid a salary, wage or earn commission.
- Additional amounts may be paid to employees in the form of overtime or bonuses.
- Employers may deduct amounts from employees' earnings such as insurance, union dues, medical aid and taxes.
- The amount a person earns before deductions is their gross earnings. The amount they actually are paid after deductions is their net earnings.

KEY CONCEPT

Calculating with money.

1. A cashier works a 38-hour week and earns $731.88. What is this as an hourly rate of pay?

2. What is the annual salary of a person who is paid $2130 per month?

3. An electrician's assistant earns $25.50 per hour for a 35-hour week and 1.5 times the hourly rate for each hour worked above 35 hours. How much would the assistant earn per week for working:

   a 36 hours   b 40 hours   c 30 hours   d $42\frac{1}{2}$ hours?

4. Sandile earns a gross salary of $1613.90 per month. The company deducts 15% income tax, $144.20 insurance and 1.5% for union dues. What is Sandile's net salary?

5. A clothing factory worker is paid the equivalent of $1.67 per completed garment. How much will the worker earn for completing 325 garments in a month?

6. A bank clerk receives an annual salary of $32 500 but is paid weekly. 4% of weekly gross earnings are paid into a pension fund and an additional $93.50 is deducted each week.

   Calculate the clerk's:

   a weekly gross earnings

   b weekly pension fund payment

   c net income per week.

TIP

Remember, work out all percentage deductions on the gross income and then subtract them all from the gross.

17.2 Borrowing and investing money

KEY LEARNING STATEMENTS

- When you borrow money you may pay interest on the amount borrowed.
- When you invest or save money you may earn interest on the amount invested.
- If the amount of interest paid (or charged) is the same for each year, then it is called simple interest.
- When the interest for one year is added to the investment (or debt) and the interest for the next year is calculated on the increased investment (or debt), it is called compound interest.
- The original amount borrowed or invested is called the principal.
- For simple interest, the interest per annum = interest rate × principal (original sum invested).
- The formula used to calculate simple interest is: $I = \dfrac{PRT}{100}$ where P is the
- principal, R is the interest rate and T is the time (in years).
- You also need to know the formula for calculating the value (V) of an investment when it is subject to compound interest:
  $V = P\left(1 + \dfrac{r}{100}\right)^n$, where P is the original amount invested, r is the rate of interest and n is the number of years of compound growth.
- Hire purchase (HP) is a method of buying things on credit and paying them off over an agreed period of time. Normally you pay a deposit and equal monthly instalments.

KEY CONCEPT

Calculating with simple and compound interest.

1. Calculate the simple interest on:
   a. $250 invested for a year at the rate of 3% per annum
   b. $400 invested for five years at the rate of 8% per annum
   c. $700 invested for two years at the rate of 15% per annum
   d. $800 invested for eight years at the rate of 7% per annum
   e. $5000 invested for 15 months at the rate of 5.5% per annum.

2. $7500 is invested at 3.5% per annum simple interest. How long will it take for the amount to reach $8812.50?

3. The total simple interest on $1600 invested for five years is $224. What is the percentage rate per annum?

4. The cash price of a car is $20 000. The hire purchase price is a $6000 deposit and instalments of $700 per month for two years. How much more than the cash price is the hire purchase price?

TIP

'per annum' (p.a.) means 'per year'

TIP

You can change the subject of the simple interest formula:

$I = \dfrac{PRT}{100}$ $P = \dfrac{100I}{RT}$

$R = \dfrac{100I}{PT}$ $T = \dfrac{100I}{PR}$

5   Lebo can pay $7999 cash for a new car or he can buy it on HP by paying a $2000 deposit and 36 monthly payments of $230. How much extra will he pay by buying on HP?

6   Calculate the compound interest on:

  a $250 invested for a year at the rate of 3% per annum

  b $400 invested for five years at the rate of 8% per annum

  c $700 invested for two years at the rate of 15% per annum

  d $800 borrowed for eight years at the rate of 7% per annum

  e $5000 borrowed for 15 months at the rate of 5.5% per annum.

7   How much will you have in the bank if you invest $500 for four years at 3% interest, compounded annually?

8   A person invests $5000 in an investment scheme for five years and earns 8% p.a. simple interest.

  a   Calculate the total interest they will earn.

  b   How much would they need to invest to earn $3600 interest in the same period (at the same rate)?

17.3 Buying and selling

KEY LEARNING STATEMENTS

- The amount a business pays for an item is called the cost price. The price they sell it for is called the selling price. The amount the seller adds onto the cost price to make a selling price is called a mark up. For example, a shopkeeper may buy an item for $2 and mark it up by 50 cents to sell it for $2.50.

- The difference between the cost price and the selling price is called the profit (if it is higher than the cost price) or the loss (if it is lower than the cost price).
    - Profit = selling price − cost price.
    - Loss = cost price − selling price.

- The percentage profit (or loss) = $\frac{\text{profit (or loss)}}{\text{cost price}} \times 100\%$.

- A discount is an intentional reduction in the price of an item.
    - Discount = original selling price − new marked price.

- Percentage discount = $\frac{\text{discount}}{\text{original selling price}} \times 100\%$.

KEY CONCEPT

Calculations involving profit and loss, including percentage calculations.

1 Find the selling price of each of the following:

   a cost price $120, profit 20%
   b cost price $230, profit 15%
   c cost price $289, loss 15%
   d cost price $600, loss $33\frac{1}{3}$%.

2 During a Shopping Festival in Dubai, a jeweller offers a 10% discount on the marked price of all items. Calculate the sale price of a necklace with a marked price of $1200.

3 If a shopkeeper buys an article for $440 and loses 12% on the sale, find the selling price.

4 A dentist offers a 5% discount to patients who pay their accounts in cash within a week. How much will someone with an account of $67.80 pay if they pay promptly in cash?

5 Calculate the new selling price of each item with the following discounts.

   a $199 discount 10%.
   b $45.50 discount 12%.
   c $1020 discount $5\frac{1}{2}$%.

REVIEW EXERCISE

1 At a rate of $19.50 per hour how many hours does a worker need to work to earn:

   a $234
   b $780
   c $497.25?

2 A mechanic works a 38-hour week for a basic wage of $28 per hour. Overtime is paid at time and a half on weekdays and double time on weekends. Calculate the gross earnings for a week if the mechanic works normal hours plus:

   a three hours overtime on Thursday
   b one extra hour per day for the whole week
   c two hours overtime on Tuesday and $1\frac{1}{2}$ hours overtime on Saturday.

3 Jamira earns a monthly salary of $5234.

   a What is her annual gross salary?
   b She pays 12% tax and has a further $456.90 deducted from her monthly salary. Calculate her net monthly income.

4 A $10 000 investment earns interest at a rate of 3% p.a. Draw a bar chart to compare the value of the investment after one, five and ten years for both types of interest. Comment on what your graph shows about the difference between simple and compound interest.

5 Find the selling price of an article that was bought for $750 and sold at a profit of 15%.

CONTINUED

6 Calculate the selling price of an item of merchandise bought for $3000 and sold at a profit of 12%.

7 A gallery owner displays paintings for artists. They add a 150% mark up on the price asked by the artist to cover expenses and make a profit. An artist supplies three paintings at the prices listed below. For each one, calculate the mark up in dollars, and the selling price the gallery owner would charge.

    a Painting A, $890.   **b** Painting B, $1300.   **c** Painting C, $12 000.

8 An art collector wants to buy paintings A and B (from question 7). They agree to pay cash on condition that the gallery owner offers a 12% discount on the selling price of the paintings.

    a What price will the collector pay?

    b What percentage profit does the gallery owner make on the sale?

9 A cyclist bought a bicycle for $500. After using it for two years, the bike was sold at a loss of 15%. Calculate the selling price.

10 An article is being sold at a loss of 12%. The cost of the article was $240. Calculate the selling price.

11 It costs a total of $377 to make ten identical dresses. At what price should each dress be sold in order to make 15% profit?

12 Sal wants to buy a used scooter. The cash price is $495. The price of the scooter on credit is a 20% deposit and then 24 monthly instalments of $25. How much will Sal save by paying cash?

SELF ASSESSMENT

Analysing your mistakes can help you improve.

Mark your own work in this Review exercise.

Consider any questions that you got wrong.

- Did you understand the question and know what to do? If not, what can you do about this?

- Did you make any errors in your calculations? If so, what can you do to avoid these errors in future?

- Would doing corrections help you? If so, redo the questions you got wrong.

Chapter 18: Curved graphs

18.1 Review of quadratic graphs (the parabola)

KEY LEARNING STATEMENTS

- Functions with an x^2 term (but no higher powers) produce a curved graph called a parabola.
- The general rule for a parabola is $y = ax^2 + bx + c$.
- The value of the coefficient of the x^2 term in the rule gives you information about the shape of the graph.
  - If a is positive, the graph is U shaped.
  - If a is negative, the graph is ∩ shaped.
- The axis of symmetry is a line with the equation $y = k$ that divides the graph into two symmetrical parts. The point where the axis of symmetry cuts the x-axis (k) is halfway between the x-intercepts of the graph.
- You can draw a table of values, plot the points and join them with a smooth curve to draw a parabola.
- To sketch a graph, use the characteristics of the graph to draw it without constructing a table of values.

KEY CONCEPTS

- Constructing a table of values to draw functions in the form $\pm x^2 + ax + b$.
- Sketching and interpreting quadratic graphs.

1. Construct a table of values, selecting values from $-6 \leqslant x \leqslant 6$, for each of the following equations. Draw the graphs on the same set of axes.

   a $y = -x^2 + 3$ **b** $y = x^2 - 2x + 2$ **c** $y = x^2 + 8x + 16$

2. Four parabolas (**A** to **D**) are shown here.

   A

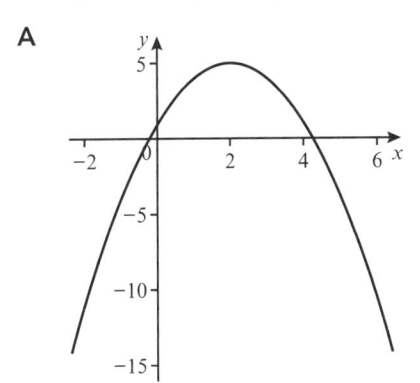

   B

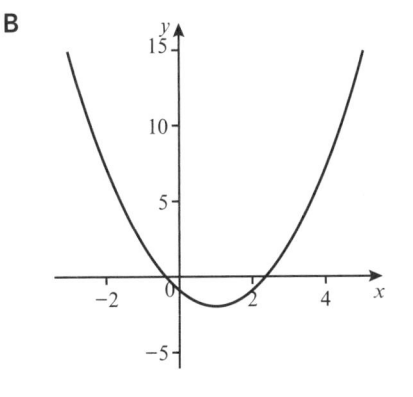

   TIP

   Remember, the constant term (c in the general formula) is the y-intercept.

C

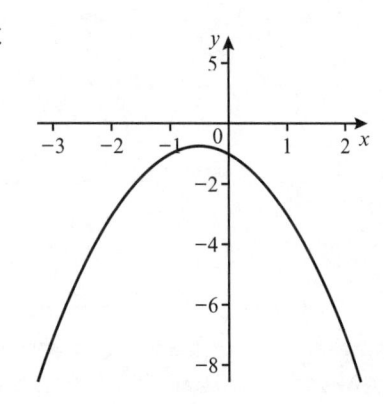

D

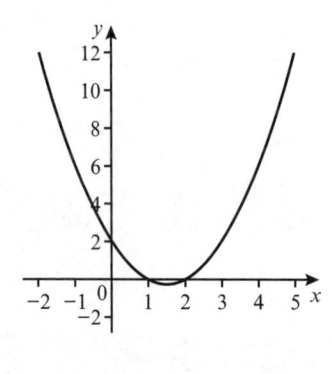

Match the equations below to the graphs. Write the letter of the graph only.

a $y = x^2 - 3x + 2$
b $y = x^2 - 2x - 1$
c $y = -x^2 + 4x + 1$
d $y = -x^2 - x - 1$

3 Draw and label sketch graphs of the following functions.

a $y = x^2 + 3x$
b $y = -2x^2 + 8$
c $y = \frac{1}{2}x^2 + 2$

4 A toy rocket is thrown up into the air.
The graph shows its path.

a What is the greatest height the rocket reaches?

b How long did it take for the rocket to reach this height?

c How high did the rocket reach in the first second?

d For how long was the rocket in the air?

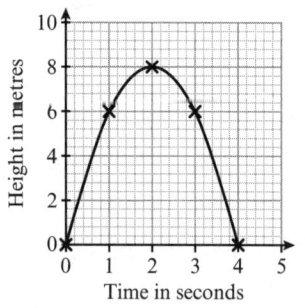

e Estimate for how long the rocket was higher than 3 metres above ground.

5 In a factory, the cost of electricity for maintaining the temperature in a cool room (y dollars) depends on the average outside temperature on that day. The relationship can be described as 'three more than double the square of the temperature (in °C)'.

a Write this rule in the form of $y = ax^2 + c$, where y is the cost in dollars and x is the temperature in °C.

b Copy and complete this table of values for temperatures from −10 to 10 °C.

x	−10	−5	0	5	10
y					

c Draw a graph of this function using the values from the table.

d Describe what happens to the cost of electricity as temperature changes.

e The cost of electricity for a day was $100. What could the average outside temperature have been? Give your answers correct to the nearest degree.

18.2 Drawing reciprocal graphs (the hyperbola)

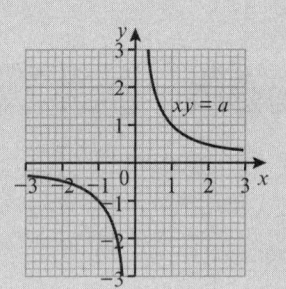

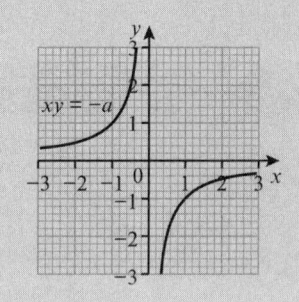

KEY CONCEPTS

- Constructing a table of values to draw functions in the form of $y = \dfrac{a}{x}$.
- Drawing and interpreting reciprocal graphs (hyperbolas).

KEY LEARNING STATEMENTS

- The graph produced by the equation $y = \dfrac{a}{x}$ or $xy = a$, where a is an integer, $x \neq 0$ and $y \neq 0$ is a reciprocal graph or hyperbola.
- The graph of $xy = a$ has two symmetrical parts drawn in diagonally opposite quadrants.
- For $xy = a$, where a is positive, the graph is in the first and third quadrants, where a is negative, the graph is in the second and fourth quadrants.

1 Copy and complete the table of values for each equation. Use the points to plot each graph on a separate set of axes.

a

x	−5	−4	−3	−2	−1	1	2	3	4	5
$y = \dfrac{2}{x}$										

b

x	−5	−4	−3	−2	−1	1	2	3	4	5
$y = -\dfrac{1}{x}$										

c

x	−5	−4	−3	−2	−1	1	2	3	4	5
$xy = 1$										

d

x	−5	−4	−3	−2	−1	1	2	3	4	5
$xy = -2$										

TIP

Reciprocal equations have a constant product. This means in $xy = a$, x and y are variables but a is a constant.

TIP

When you work out the values, you can round one or two decimal places as this is usually accurate enough for you to draw the graph.

2   The length and width of a certain rectangle can only be a whole number of metres. The area of the rectangle is 24 m².

   a   Draw a table that shows all the possible combinations of measurements for the length and width of the rectangle.

   b   Plot your values from part (a) as points on a graph.

   c   Join the points with a smooth curve. What does this graph represent?

   d   Assuming, now, that the length and width of the rectangle can take any positive values that give an area of 24 m², use your graph to find the width if the length is 7 metres.

18.3 Using graphs to solve quadratic equations

KEY LEARNING STATEMENTS

- If a quadratic equation has real roots the graph of the equation will intersect with the x-axis. This is where $y = 0$.
- To solve quadratic equations graphically, read off the x-coordinates of the points for a given y-value.

KEY CONCEPT

Using graphs to find or estimate solutions to equations.

1   This is the graph of $y = x^2 - 4x + 3$. Use the graph to estimate the solution of:

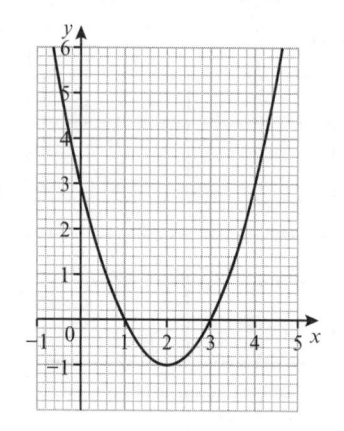

TIP

For part (c) you might find it helpful to rearrange the equation so the left-hand side matches the equation of the graph, i.e. add three to both sides.

   a $x^2 - 4x + 3 = 0$ b $x^2 - 4x + 3 = 3$ c $x^2 - 4x = 1$

2   a   Draw the graph of $y = x^2 - 4x - 5$, for x-values from -2 to 6.

   b   Use the graph to find the approximate solutions of the equations:

      i $x^2 - 4x - 5 = 0$

      ii $x^2 - 4x - 5 = -8$

      iii $x^2 - 4x = 2$

3 a Use an interval of $-4 \leq x \leq 5$ on the x-axis to draw the graph $y = x^2 - x - 6$.

   b Use the graph to solve the following equations:

   i $-6 = x^2 - x - 6$ ii $x^2 - x - 6 = 0$

   iii $x^2 - x = 12$

SELF ASSESSMENT

Complete the Review exercise and mark your own work.

Can you answer yes to all of these questions?

If not, write down what you need to do to change your answer.

- Did I understand all of the questions?
- Did I know what to do to answer them?
- Did I follow the instructions correctly?
- Was my work clear and organised?
- Did I perform procedures accurately?
- Were my solutions correct?
- If I made mistakes, could I work out what I did wrong?

REVIEW EXERCISE

1 Look at these sketch graphs. For each one, write its equation.

a [parabola through $(-2, 0)$ and $(2, 0)$ with minimum at -4]

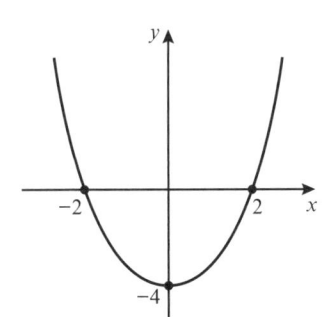

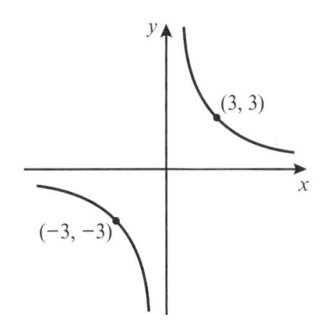

b [hyperbola through $(3, 3)$ and $(-3, -3)$]

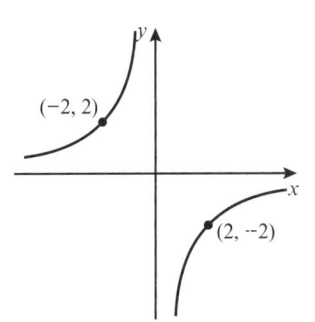

c [curve through $(-2, 2)$ and $(2, -2)$]

d [parabola with maximum at 9, roots at -3 and 3]

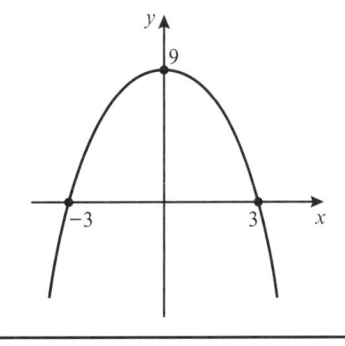

CONTINUED

2 Sketch the following graphs. Label the graphs and any intercepts with the axes.

    a $y = 3x^2$ **b** $y = -x^2 + 4$ **c** $y = \dfrac{12}{x}$

3 Look at these two graphs.

A

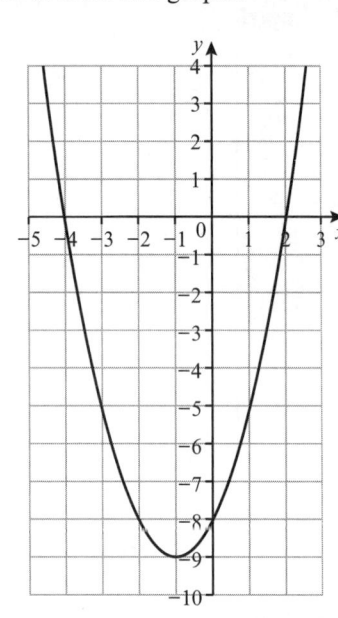

B

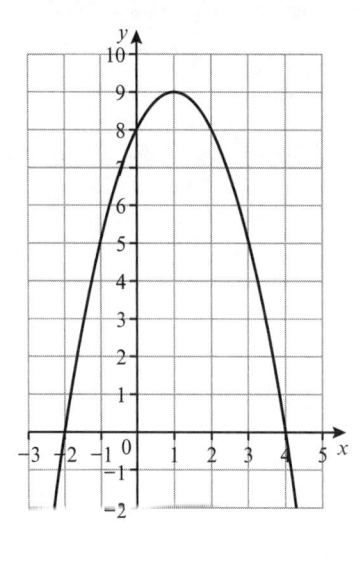

  a The equations of the graphs are $y = x^2 + 2x - 8$ and $y = -x^2 + 2x + 8$. Explain how you can tell which graph is which.

  b Use the appropriate graph to solve each of these equations.

      i $x^2 + 2x - 8 = 0$ **ii** $x^2 = 2x + 8$ **iii** $x^2 + 2x - 8 = -5$

      iv $-x^2 + 2x + 8 = 9$ **v** $-x^2 + 2x = -5$

Chapter 19: Symmetry

19.1 Symmetry in two dimensions

> **KEY LEARNING STATEMENTS**
>
> - Two-dimensional (flat) shapes have line symmetry if you are able to draw a line through the shape so that one side of the line is the mirror image (reflection) of the other side. There may be more than one possible line of symmetry in a shape.
>
> - If you rotate (turn) a shape around a fixed point and it fits on to itself during the rotation, then it has rotational symmetry. The number of times the shape fits on to its original position during a 360° rotation is called the order of rotational symmetry.

> **KEY CONCEPT**
>
> Line and rotational symmetry in 2D shapes.

1. **a** How many lines of symmetry are there for each letter in this word in the font shown here?

    CAMBRIDGE

    b Which letters in the word have rotational symmetry and what is the order of rotation?

2. For each of the following shapes:

    a copy the shape and draw in any lines of symmetry

    b determine the order of rotational symmetry.

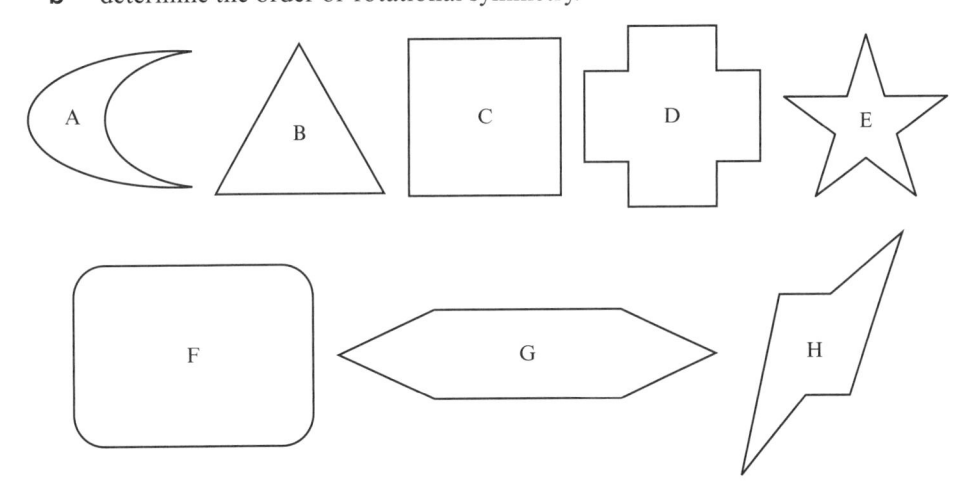

> **TIP**
>
> If a shape can only fit back onto itself after a full 360° rotation, it has no rotational symmetry.

3. **a** How many lines of symmetry does a rhombus have? Draw a diagram to show your solution.

    b What is the order of rotational symmetry of a rhombus?

4. Draw a quadrilateral that has no lines of symmetry and no rotational symmetry.

19.2 Angle relationships in circles

KEY LEARNING STATEMENTS

- When a triangle is drawn in a semicircle, so that one side is the diameter and the vertex opposite the diameter touches the circumference, the angle of the vertex opposite the diameter is a right angle (90°).

- Where a tangent touches a circle, the radius drawn to the same point meets the tangent at 90°.

KEY CONCEPT

Angle relationships in circles.

1 Determine the size of angle x in each diagram. Show your working and give reasons for any statements you make.

a, b, 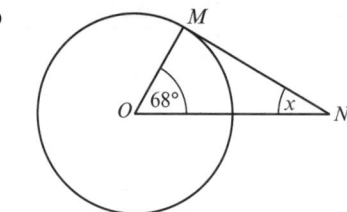 c, 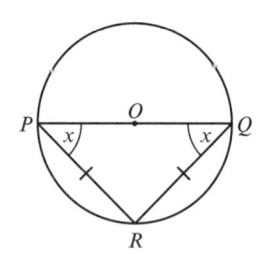 d, 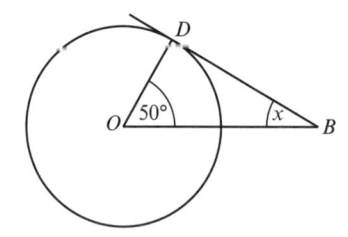 e, 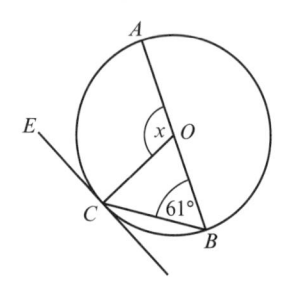 f

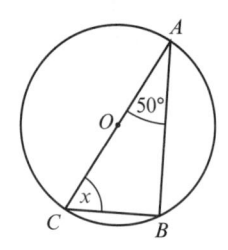

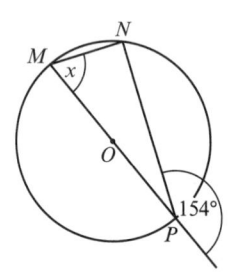

2 Given that O is the centre of the circle, calculate the value of angle BAD with reasons.

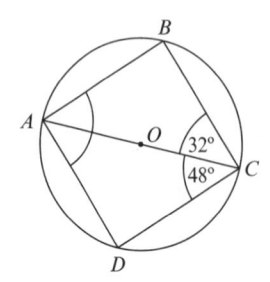

19 Symmetry

3 DC is a tangent to the circle with centre O. Find the size of angle DCA.

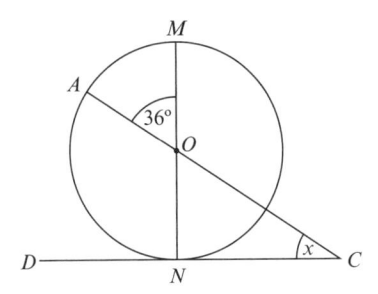

4 EC is a tangent to the circle with centre O. AB is a straight line and angle CBD = 37°. Calculate the size of the angles marked w, x, y and z giving reasons for each.

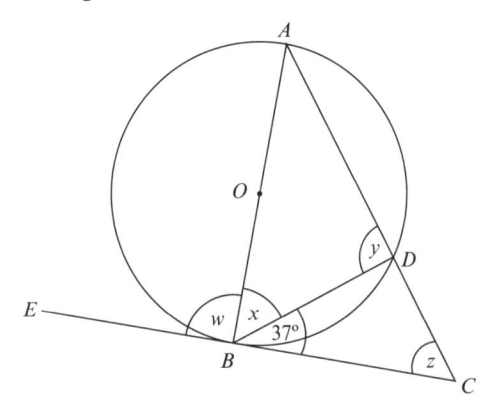

REVIEW EXERCISE

1 Here are five shapes.

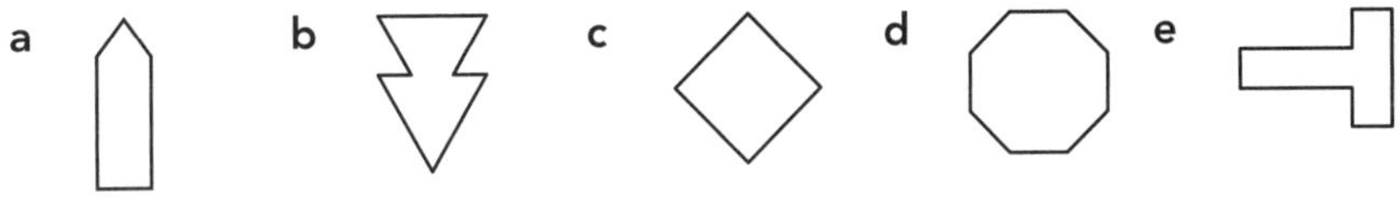

a b c d e

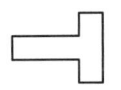

For each one state:

i the number of lines of symmetry (if any)

ii the order of rotational symmetry.

CONTINUED

2 Find the sizes of angles *a* to *e* in the diagram.

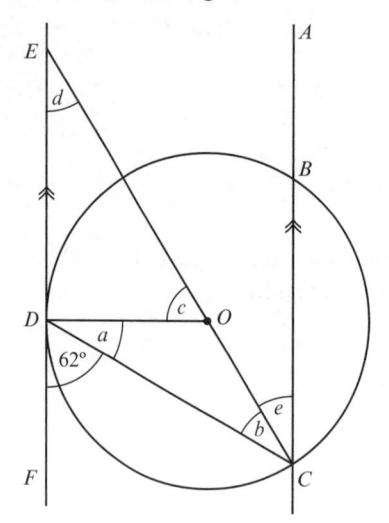

TIP

The angle relationships for triangles, quadrilaterals and parallel lines (Chapter 3), as well as Pythagoras' theorem (Chapter 11), may be needed to solve circle problems.

SELF ASSESSMENT

How well do you understand the work on circles?

1. Look back over your completed exercise.

2. Complete these sentences to summarise what went well:

    a The things that went well were …

    b The best bit of my work was …

3. How could you improve your work? Complete these sentences about your own work:

    a To improve my work I need to …

    b Next time I solve problems involving angles in circles I must remember to …

    c I think I could improve if I focussed on …

Chapter 20: Ratio, rate and proportion

20.1 Working with ratio

KEY LEARNING STATEMENTS

- A ratio is a comparison of two or more quantities measured in the same units. In general, a ratio is written in the form $a:b$.
- You should always give ratios in their simplest form. To express a ratio in simplest form, divide or multiply all parts of the ratio by the same number.
- Quantities can be shared in a given ratio. To do this you work out the number of equal parts in the ratio and then work out the value of each share. For example, a ratio of $3:2$ means that there are five equal parts. One share is $\frac{3}{5}$ of the total and the other is $\frac{2}{5}$ of the total.

KEY CONCEPTS

- Simplifying ratios.
- Dividing quantities in a given ratio.

1. Express the following as ratios in their simplest form.

    a $120:150$
    b $2\frac{3}{4}:3\frac{2}{3}$
    c $1\frac{1}{2}$ hours : 15 minutes
    d   125 litres to 350 litres
    e   45 cents to $2
    f   175 cm to 2 metres
    g   600 g to three kilograms
    h   50 mm to a metre
    i   12.5 g to 50 g
    j   3 cm to 25 mm
    k   200 ml to 3 litres

    TIP
    Remember, simplest form is also called 'lowest terms'.

2. Find the value of x in each of the following.

    a $2:3 = 6:x$
    b $2:5 = x:10$
    c $2:x = 3:24$
    d $x:12 = 2:8$
    e $10:15 = x:6$
    f $\frac{2}{7} = \frac{x}{4}$
    g $\frac{5}{x} = \frac{16}{6}$
    h $\frac{x}{4} = \frac{10}{15}$
    i $\frac{x}{21} = \frac{1}{3}$
    j $\frac{5}{x} = \frac{3}{8}$
    k $\frac{3}{35} = \frac{x}{28}$
    l $0.8:20 = x:4$

    TIP
    You can cross multiply to make an equation and solve for x.

3. Two people shared $12 000 in the ratio $3:2$. Calculate each person's share.

4. Divide 350 sheets of paper in the ratio $2:5$.

5. A length of rope 160 cm long must be cut into two parts so that the lengths are in the ratio $3:5$. What are the lengths of the parts?

6. To make salad dressing, you mix oil and vinegar in the ratio $2:3$. Calculate how much oil and how much vinegar you will need to make the following amounts of salad dressing:

    a   50 ml
    b   600 ml
    c   750 ml.

7   The sizes of three angles of a triangle are in the ratio $A:B:C = 2:1:3$.
    What is the size of each angle?

8   A metal disc consists of three parts silver and two parts copper (by mass). If the disc has a mass of 1350 mg, how much silver does it contain?

20.2 Ratio and scale

KEY LEARNING STATEMENTS

- Scale is a ratio. It can be expressed as 'length on the drawing : real length'.
- All ratio scales must be expressed in the form $1:n$ or $n:1$.
- To change a ratio so that one part = 1, you divide both parts by the number that you want to be expressed as 1. For example with $2:7$, if you want the two to be expressed as one, you divide both parts by two. The result is $1:3.5$.

KEY CONCEPT

Using ratio to solve scale problems.

1   Write these ratios in the form of $1:n$.

    a $4:9$ b 400 metres : 1.3 km   c 50 minutes : $1\frac{1}{2}$ hours

2   Write these ratios in the form of $n:1$.

    a $12:8$ b 2 metres : 40 cm   c 2.5 g to 500 mg

3   The distance between two points on a map with a scale of $1:2\,000\,000$ is 120 mm. What is the distance between the two points in real life? Give your answer in kilometres.

4   A plan is drawn using a scale of $1:500$. If the length of a wall on the plan is 6 cm, how long is the real wall?

5   Miguel makes a scale drawing to solve a trigonometry problem. 1 cm on his drawing represents 2 metres in real life. He wants to show a 10 metre long ladder placed 7 metres from the foot of a wall.

    a   What length will the ladder be in the diagram?

    b   How far will it be from the foot of the wall in the diagram?

6   A map has a scale of $1:700\,000$.

    a   What does a scale of $1:700\,000$ mean?

    b   Copy and complete this table using the map scale.

Map distance (mm)	10		50	80		
Actual distance (km)		50			1200	1500

TIP

With reductions (such as maps) the scale will be in the form $1:n$, where $n > 1$. With enlargements the scale will be in the form $n:1$, where $n > 1$. n may not be a whole number.

7   This plan of a botanical garden is drawn at a scale of 1 : 750.

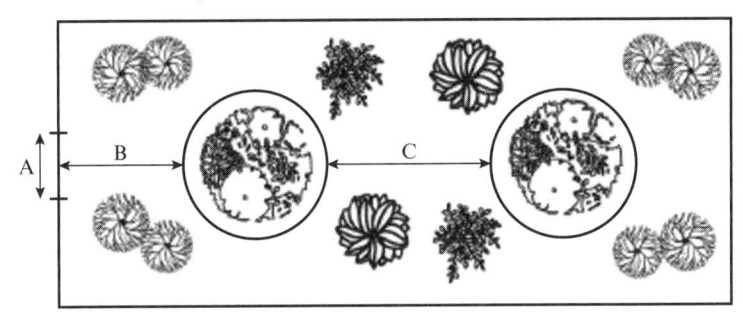

   a   Measure the lengths *A*, *B* and *C* shown on the diagram and write them in mm.

   b   Use the scale to work out the actual lengths *A*, *B* and *C*.

8   A designer has a rectangular picture 35 mm wide and 37 mm high. They enlarge it using a photocopier so that the enlargement is 14 cm wide.

   a   What is the scale factor of the enlargement?

   b   What is the height of her enlarged picture?

   c   In the original picture, a fence was 30 mm long. How long will this fence be on the enlarged picture?

TIP

An enlargement of the picture is the 'drawing' or 'map' in this example.

20.3 Rates

KEY LEARNING STATEMENTS

- A rate compares two quantities measured in different units. For example, speed is a rate that compares kilometres travelled per hour.
- Rates can be simplified just like ratios.
- You solve rate problems in the same way that you solved ratio and proportion problems. Use the unitary or ratio methods.

KEY CONCEPT

Using and applying measures of rate.

TIP

The word 'per' is often used in a rates. Per can mean 'for every', 'in each', 'out of every', or 'out of' depending on the context.

1   Milk costs $1.95 per litre. How much milk can you buy for $50?

2   Sam travels a distance of 437 km and uses 38 litres of petrol. Express his petrol consumption as a rate in kilometres per litre .

> **TIP**
>
> Remember, speed is a very important rate.
>
> speed = $\frac{\text{distance}}{\text{time}}$
>
> distance = speed × time
>
> time = $\frac{\text{distance}}{\text{speed}}$

3 Calculate the average speed of the following vehicles.

    a A car that travels 196 km in 2.5 hours.

    b A plane that travels 650 km in one hour 15 minutes.

    c A train that travels 180 km in 45 minutes.

4 How long will it take to travel:

    a 400 km at 80 km/h   **b** 900 km at 95 km/h

    c 1800 km at 45 km/h   **d** 500 metres at 7 km/h.

5 How far will you travel in $2\frac{1}{2}$ hours at these speeds?

    a 60 km/h   **b** 120 km/h

    c 25 metres per minute   **d** two metres per second

6 Water runs out of a pipe at a rate of 0.15 litres per second. How long will it take to fill a 25 litre container?

7 A gold bar has a volume of 725 cm³ and a mass of 14.5 kg.

    Given that density = $\frac{\text{mass}}{\text{volume}}$, calculate the density of the gold bar in g/cm³.

20.4 Kinematic graphs

> **KEY LEARNING STATEMENTS**
>
> - On a distance–time graph the slope (steepness) of the line shows speed.
>
>   Speed = $\frac{\text{distance travelled}}{\text{time taken}} = \frac{\text{change in } y\text{-coordinate}}{\text{change in } x\text{-coordinate}}$
>
>   A straight line indicates constant speed; the steeper the line, the greater the speed; and a horizontal line represents no movement.
>
> - The gradient on a distance–time graph gives the speed *and* direction of motion (this is the velocity). So, upward and downward slopes represent movement in opposite directions.

> **KEY CONCEPTS**
>
> - Solving problems involving average speed.
> - Drawing and interpreting kinematic graphs.
> - Graphs in practical situations.

20 Ratio, rate and proportion

1 The graph below shows the distance covered by a vehicle in a six-hour period.

   a Use the graph to find the distance covered after:

      i one hour

      ii two hours

      iii three hours.

   b Calculate the speed of the vehicle during the first three hours.

   c Describe what the graph shows between hour three and four.

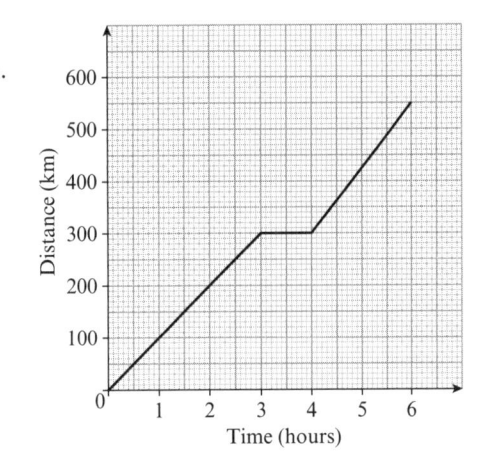

   d What distance did the vehicle cover during the last two hours of the journey?

   e What was its speed during the last two hours of the journey?

2 Dabilo and Pam live 200 km apart from each other. They decide to meet up at a shopping centre in-between their homes. Pam travels by bus and Dabilo catches a train. The graph shows both journeys.

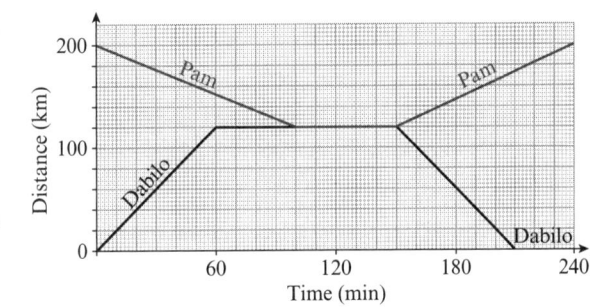

   a How much time did Dabilo spend on the train?

   b How much time did Pam spend on the bus?

   c At what speed did the train travel for the first hour?

   d How far was the shopping centre from:

      i Dabilo's home?   **ii** Pam's home?

   e What was the speed of the bus from Pam's home to the shopping centre?

   f How long did Dabilo have to wait before Pam arrived?

   g How long did the Pam and Dabilo spend together?

   h How much faster was Pam's journey on the way home?

   i If they left home at 8:00 a.m., what time did each person return home after the day's outing?

3   At break time, Ani leaves the classroom and takes 10 seconds to walk 25 metres to the tuck shop. Ani waits 25 seconds for a drink and then takes 5 seconds to walk a further 7 metres to the bin before turning and taking the same route back to the classroom, taking 10 seconds to get there.

   a   Draw a distance time graph to show how Ani's distance from the classroom changed over time.

   b   What was Ani's speed for the first ten seconds?

   c   Calculate the total distance that Ani walked.

   d   Use the distance–speed–time formula to calculate Ani's average speed for the whole journey.

20.5 Proportion

KEY LEARNING STATEMENTS

- Proportion compares a part with the whole. It is usually expressed as a fraction, percentage or ratio.
- When quantities are in direct proportion they increase or decrease at the same rate. The graph of a directly proportionate relationship is a straight line passing through the origin.
- When quantities are inversely proportional, one increases as the other decreases. The graph of an inversely proportional relationship is a curve.
- The unitary method is useful for solving ratio and proportion problems. This method involves finding the value of one unit (of work, time, etc.) and using that value to find the value of a number of units.

KEY CONCEPT

Direct and inversely proportional relationships.

TIP

If x and y are directly proportional, then $\frac{x}{y}$ is the same for various values of x and y.

1   Determine whether A and B are directly proportional in each case.

   a
A	2	4	6
B	300	600	900

   b
A	2	5	8
B	2	10	15

   c
A	1	2	3	4
B	0.1	0.2	0.3	0.4

2   A textbook costs $25.

   a   What is the price of seven books?   b   What is the price of ten books?

3   Find the cost of five identically priced items if seven items cost $17.50.

4   If a 3.5 metre tall pole casts a 10.5 metre shadow, find the length of the shadow cast by a 20 metre tall pole at the same time.

5   A car travels a distance of 225 km in three hours at a constant speed.

   a   What distance will it cover in one hour at the same speed?

   b   How far will it travel in five hours at the same speed?

   c   How long will it take to travel 250 km at the same speed?

6   A truck uses 20 litres of fuel to travel 240 km.

   a   How much fuel will it use to travel 180 km at the same rate?

   b   How far can the truck travel on 45 litres of fuel at the same rate?

7   It takes one employee ten days to complete a project. If another employee joins him, it only takes five days. Five employees can complete the job in two days.

   a   Describe this relationship.

   b   How long would it take to complete the project with:

       i   four employees   ii   20 employees?

8   At a campsite, ten people have enough fresh water to last them for six days at a set rate per person.

   a   How long would the water last, if there were only five people drinking it at the same rate?

   b   Another two people join the group. How long will the water last if it is used at the same rate?

9   Mia took four hours to complete a journey at 110 km/h. Huan did the same journey at 80 km/h. How long did it take Huan?

10 A plane travelling at an average speed of 1000 km/h takes 12 hours to complete a journey. How fast would it need to travel to cover the same distance in ten hours?

SELF ASSESSMENT

Copy the headings and complete a learning log for the work on direct and inverse proportion.

Learning log for:	Direct and inverse proportion
I learned …	
I still need to learn more about ..	
My next steps will be …	
I need some help with …	
In general, I feel _____ about this section of work.	

REVIEW EXERCISE

1. Express the following as ratios in their simplest form.

   a $3\frac{1}{2} : 4\frac{3}{4}$ b 5 ml to 2.5 litres   c 125 g to 1 kg

2. Divide 600 in these ratios.

   a 7 : 3   b 7 : 5   c 7 : 13   d 7 : 7

3. A builder mixes sand and cement in the ratio 4 : 1 to make mortar for a wall. How much cement will be needed for the following amounts of sand:

   a four spadefuls   b two bags

   c $1\frac{1}{2}$ wheelbarrows full?

4. A triangle of perimeter 360 mm has side lengths in the ratio 3 : 5 : 4.

   a Find the lengths of the sides.

   b Is the triangle right angled? Give a reason for your answer.

5. A model of a car is built to a scale of 1 : 50. If the real car is 2.5 metres long, what is the length of the model in centimetres?

6. On a floor plan of a school, 2 cm represents 1 m in the real school. What is the scale of the plan?

7. An athlete runs 100 metres in 9.9 seconds. Express this speed in:

   a metres per second   b kilometres per hour.

8. A car travels at an average speed of 85 km/h.

   a What distance will the car travel in:

   i 1 hour   ii $4\frac{1}{2}$ hours   iii 15 minutes?

   b How long will it take the car to travel:

   i 30 km   ii 400 km   iii 100 km?

CONTINUED

9 This travel graph shows the journey of a petrol tanker doing deliveries.

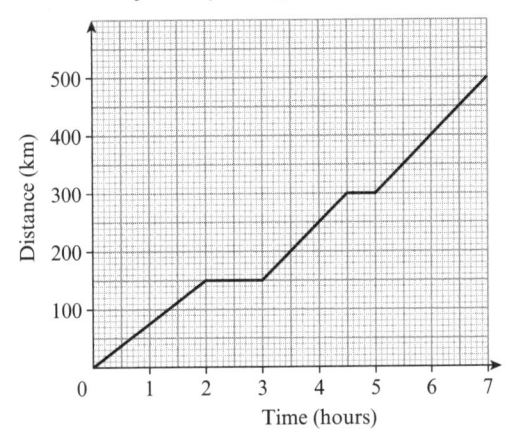

    a What distance did the tanker travel in the first two hours?

    b When did the tanker stop to make its first delivery? For how long did it stop?

    c Calculate the speed of the tanker between the first and second stop on the route.

    d What was the speed of the tanker during the last two hours of the journey?

    e How far did the tanker travel on this journey?

10 Nine students complete a task in three minutes. How long would it take six students to complete the same task if they worked at the same rate?

11 A cube with sides of 2 cm has a mass of 12 g. Find the mass of another cube made of the same material if it has sides of 5 cm.

Chapter 21: More equations and formulae

21.1 Setting up equations to solve problems

> **KEY LEARNING STATEMENTS**
>
> - You can set up your own equations and use them to solve problems.
> - The first step in setting up an equation is to work out what needs to be calculated. Represent this amount using a variable (usually x). Then construct an equation using the information you are given and solve it to find the answer.

KEY CONCEPT

Setting up equations to represent and solve problems.

1. In each of the following, set up an equation and solve it to find the unknown value.

    a   When five is added to a number, the result is 14. Find the number.

    b   If a number is decreased by seven the answer is 19. What is the number?

    c   Twice a certain number increased by 12 is 280. What is the number?

    d   The product of six and a certain number is equal to four times the number plus 16. What is the number?

    e   A third of a number is equal to half of the same number *after* it has been decreased by two. Determine the number.

    f   Half of a number is subtracted from four times itself to get seven. What is the number?

    g   The difference between two numbers is five. Three times the greater number is equal to ten times the smaller plus one. Find the numbers.

    h   The sum of two numbers is 100. Twice one number is equal to the other number plus 20. Find the numbers.

2. A rectangle has an area of $36\,\text{cm}^2$ and the width is $4\,\text{cm}$. Find the length of the rectangle.

3. A rectangle is $4\,\text{cm}$ longer than it is wide. If the rectangle is $x\,\text{cm}$ long, write down:

    a   the width (in terms of x)

    b   a formula for calculating the perimeter (P) of the rectangle.

    c   a formula for finding the area (A) of the rectangle in simplest terms.

TIP

For some problems you might need to set up two equations and solve them simultaneously to find the solutions.

TIP

When you have to give a formula, you should express it in simplest terms by collecting like terms.

4   Three numbers are represented by x, $3x$ and $x + 2$.

   a   Write a formula for finding the sum (S) of the three numbers.

   b   Write a formula for finding the mean (M) of the three numbers.

5   A triangle has sides of x cm, $x + 4$ cm and $x + 8$ cm.

   a   Write a formula in terms of x for calculating P, the perimeter of the triangle.

   b   Use your formula to find the lengths of each side of a triangle when:

      i $P = 45$ cm   ii $P = 23.25$ cm.

6   The smallest of three consecutive numbers is x.

   a   Express the other two numbers in terms of x.

   b   Write a formula for finding the sum (S) of the numbers.

7   Salma is two years older than Max. Tayo is three years younger than Max.
    If Max is x years old, write down:

   a   Salma's age in terms of x

   b   Tayo's age in term of x

   c   a formula for finding the combined ages of the three people.

8   There are 12 more apples than oranges in a bag containing 40 pieces of fruit. How many oranges are there?

9   The perimeter of a parallelogram is 104 cm. If the length is three times the width, calculate the dimensions of the parallelogram.

10 Jess thinks of a number. When she doubles it and then adds five, she gets the same answer as when she subtracts the number from two.

   a   Make an equation to represent this problem. Let the unknown number be x.

   b   Solve your equation to find the number that Jess was thinking of.

11 There are ten times more silver cars than red cars in a parking area. If there 88 silver and red cars altogether, find the number of cars of each colour.

12 Nadira is 25 years younger than her father. Nadira's mother is two years younger than her father. Together Nadira, her mother and father have a combined age of 78. Work out their ages.

REFLECTION

Have the activities in this section improved your skills in writing equations to solve problems? How?

Are there are ideas that you are still unsure of? Which?

What can you do to improve in these areas?

REVIEW EXERCISE

1. A certain number increased by six is equal to twice the same number decreased by four. What is the number?

2. If three times a number is increased by five, the result is 17. What is the number?

3. Three times a number decreased by two is equal to the sum of the number and six. Determine the number.

4. Faisal has $16 more than Nathi. Together they have $150. How much does each person have?

5. A vendor has a certain number of cakes to sell. All except 15 sell for $6 each and the vendor receives $240. How many cakes did the vendor have to sell?

6. Two items together cost $60. If a discount of 25% is given on one item, then the two items together cost $50. What was the original price of each item?

7. A rectangle of area A has sides of $x + 3$ and $x - 2$.

    a Write a formula in terms of x for finding P, the perimeter of the rectangle.

    b Find the length of each side if the rectangle has a perimeter of 98.

8. Greenburg is located between Brownburg and Townburg. Greenburg is five times as far away from Townburg as it is from Brownburg.

    If the distance between Brownburg and Townburg is 864 km, how far is it from Brownburg to Greenburg?

9. Amira is twice as old as her cousin Pam. Seven years ago, their combined age was 19.

    What are their present ages?

10. Javi left town A to travel to town B at 6.00 a.m. Town B is at least 900 km away from town A. He drove at an average speed of 90 km/h. At 8.30 a.m., Shumi left town A to travel to town B. He drove at an average speed of 120 km/h. At what time will Shumi catch up with Javi?

11. A journey took 50 minutes to complete. The driver travelled half the distance at a speed of 120 km/h and the other half at 80 km/h. How far was the journey?

Chapter 22: Transformations

22.1 Simple plane transformations

KEY LEARNING STATEMENTS

- A transformation is a change in the position or size of a point or a shape.
  - The original shape is called the object (O) and the transformed shape is called the image (O').
- There are four basic transformations: reflection, rotation, translation and enlargement.
- A reflection flips the shape over.
  - Under reflection, every point on a shape is reflected in a mirror line to produce a mirror-image of the object. Points on the object and corresponding points on the image are the same perpendicular distance from the mirror line.
  - To describe a reflection you need to give the equation of the mirror line.
- A rotation turns the shape around a point.
  - The point about which a shape is rotated is called the centre of rotation. The shape may be rotated clockwise or anti-clockwise.
  - To describe a rotation you need to give the centre of rotation and the angle and direction of turn.
- A translation is a slide movement.
  - Under translation, every point on the object moves the same distance in the same direction to form the image. Translation involves moving the shape sideways and/or up and down.
  - To describe the translation, use a column vector $\begin{pmatrix} x \\ y \end{pmatrix}$ where x is the movement to the side (along the x-axis) and y is the movement up or down. The sign of the x or y value gives you the direction of the translation. Positive means to the right or up and negative means to the left or down.

KEY CONCEPTS

- Reflection, rotation, enlargement and translation of shapes.
- Using vectors to describe translations.

CAMBRIDGE IGCSE™ MATHEMATICS: CORE PRACTICE BOOK

CONTINUED

- An enlargement involves changing the size of an object to produce an image that is similar in shape to the object.
  - The scale factor = $\dfrac{\text{length of a side on the image}}{\text{length of the corresponding side on the object}}$.

  When an object is enlarged from a fixed point, it has a centre of enlargement. The centre of enlargement determines the position of the image. Lines drawn through corresponding points on the object and the image will meet at the centre of enlargement.

  - To describe an enlargement you need to give the centre of enlargement and the enlargement factor.

You will need squared paper for this exercise.

1 Draw and label any rectangle $ABCD$.

   a Rotate the rectangle clockwise 90° about point D. Label the image $A'B'C'D'$.

   b Reflect $A'B'C'D'$ about $C'D'$.

2 Make a copy of the diagram and carry out the following transformations.

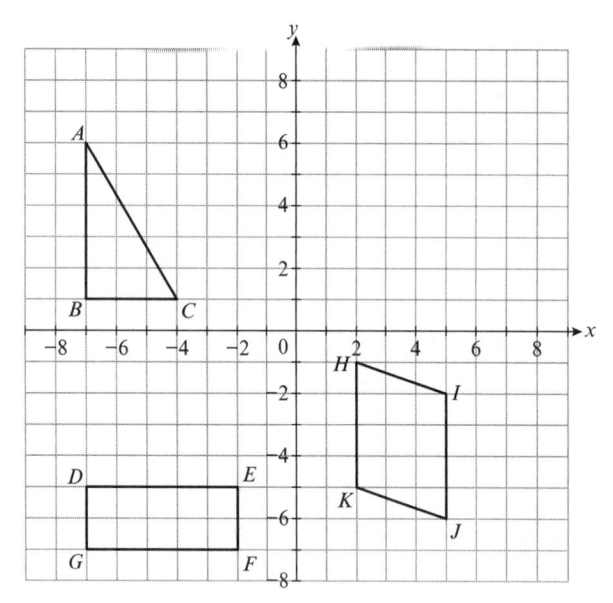

   a Translate triangle ABC three units to the right and four units up. Label the image correctly.

   b Reflect rectangle $DEFG$ about the line $y = -3$. Label the image correctly.

   c i Rotate parallelogram $HIJK$ 90° anticlockwise about point $(2, -1)$.

      ii Translate the image $H'I'J'K'$ one unit left and five units up.

TIP

Reflection and rotation change the position and orientation of the object while translation only changes the position. Enlargement changes the size of the object to produce the image.

3 Make a copy of the diagram and carry out the following transformations.

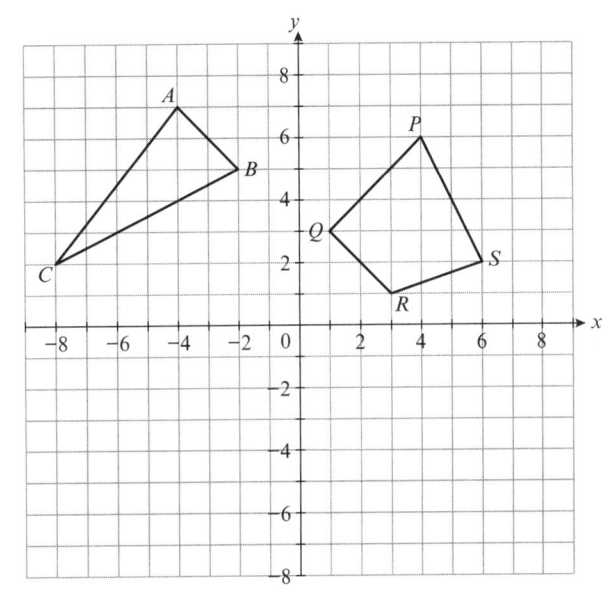

a Translate triangle ABC using the column vector $\begin{pmatrix} 10 \\ -9 \end{pmatrix}$ to form the image $A'B'C'$.

b **i** Reflect quadrilateral $PQRS$ in the y-axis.

ii Translate the image $P'Q'R'S'$ using the column vector $\begin{pmatrix} 0 \\ -6 \end{pmatrix}$.

Label the resultant image $P''Q''R''S''$.

4 For each of the reflections shown in the diagram, give the equation of the mirror line.

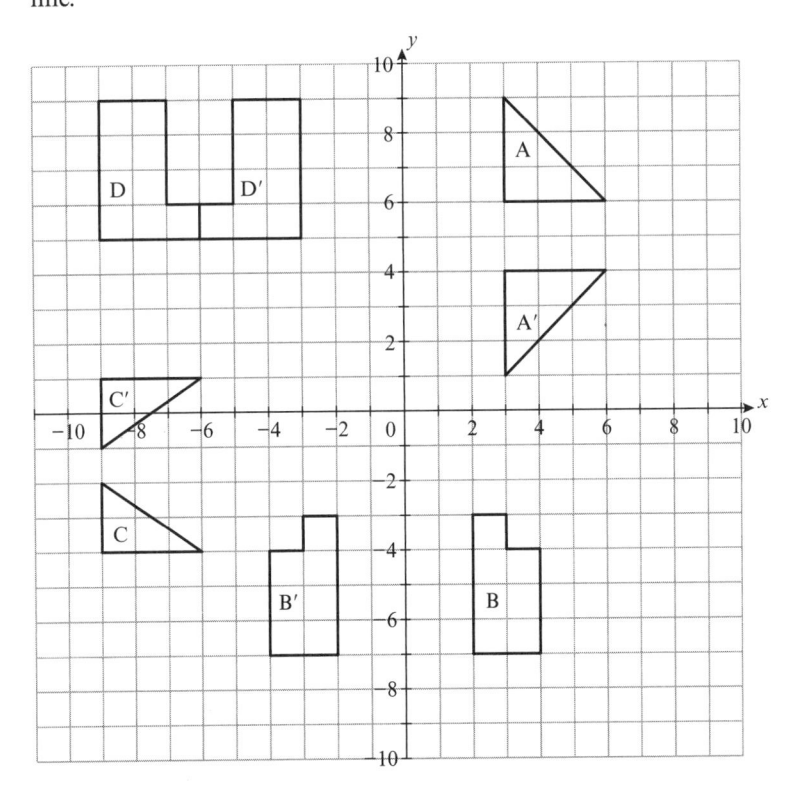

5 Copy the diagram showing shapes A–D from question 4 and draw the reflection of each shape in the *x*-axis.

6 For each pair of shapes, give the coordinates of the centre of enlargement and the scale factor of the enlargement.

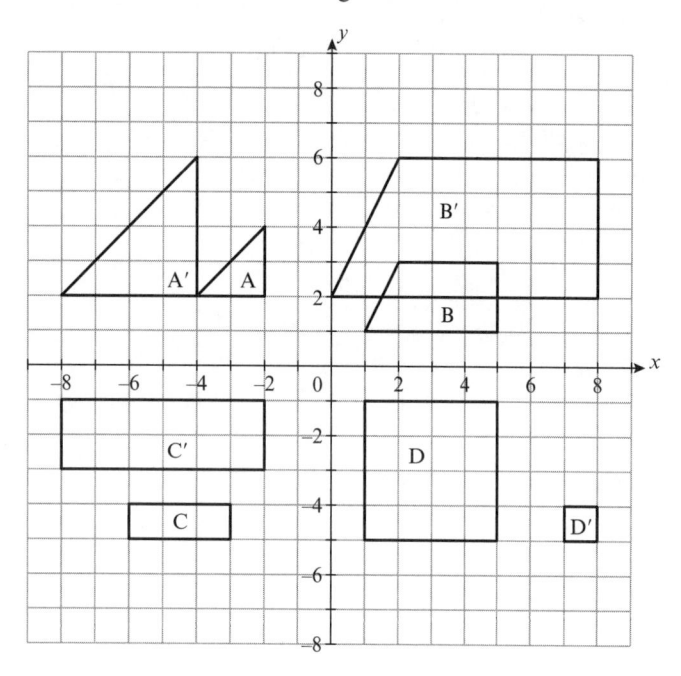

7 Copy each shape onto squared grid paper. Using point *X* as a centre of enlargement, draw the image of each shape under an enlargement with scale factor:

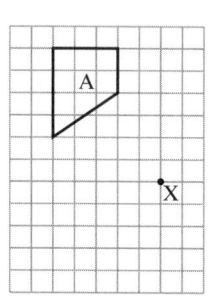

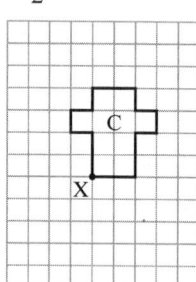

   i 2

   ii $\frac{1}{2}$

8 Triangle *ABC* is to be reflected in the *y*-axis and its image triangle *A'B'C'* is then to be reflected in the *x*-axis to form triangle *A"B"C"*.

   a Draw a set of axes, extending them into the negative direction. Copy triangle *ABC* onto your grid and draw the transformations described.

   b Describe the single transformation that maps triangle *ABC* directly onto triangle *A"B"C"*.

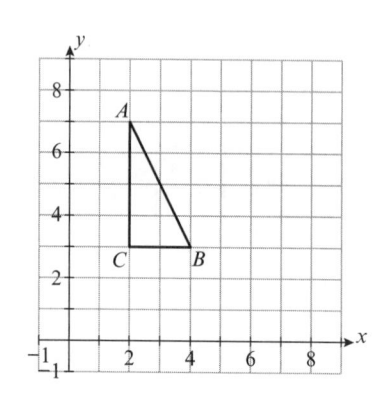

> **TIP**
>
> You can find the centre of enlargement by drawing lines through the corresponding vertices on the two shapes. The lines will meet at the centre of enlargement.
>
> When the image is smaller than the object, the scale factor of the 'enlargement' will be a fraction.

9 Shape A is to be enlarged by a scale factor of two, using the origin as the centre of enlargement, to get shape B. Shape B is then translated, using the column vector $\begin{pmatrix} -8 \\ 1 \end{pmatrix}$, to get shape C.

   a Draw a set of axes, extending the x-axis into the negative direction. Copy shape A onto your grid and draw the two transformations described.

   b What single transformation would have the same results as these two transformations?

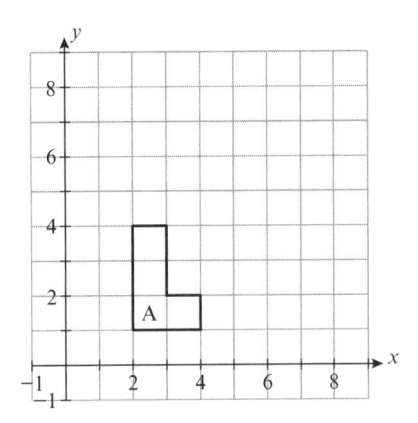

REVIEW EXERCISE

1

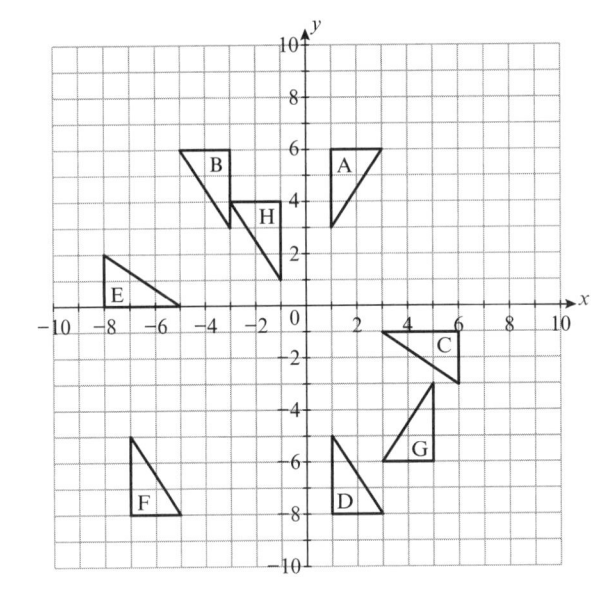

   a Describe a single transformation that maps triangle A onto:

      i   triangle B
      ii triangle C
      iii triangle D.

   b Describe the pair of transformations that you could use to map triangle A onto:

      i   triangle E
      ii triangle F
      iii triangle G
      iv triangle H.

CONTINUED

2 Sally translated parallelogram *DEFG* along the column vector $\begin{pmatrix} -4 \\ 5 \end{pmatrix}$ and then rotated it 90° clockwise about the origin to get the image *D'E'F'G'* shown on the grid.

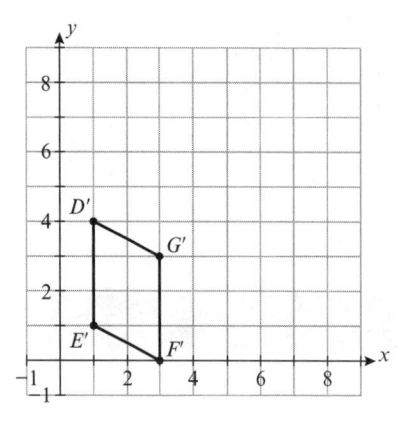

a Draw a diagram and reverse the transformations Sally performed on the shape to show the original position of *DEFG*.

b Enlarge *DEFG* by a scale factor of 2 using the origin as the centre of enlargement. Label your enlargement *D"E"F"G"*.

3 Square *ABCD* is shown on the grid.

Onto a copy of the diagram, draw the following transformations and, in each case, give the coordinates of the new position of vertex *B*.

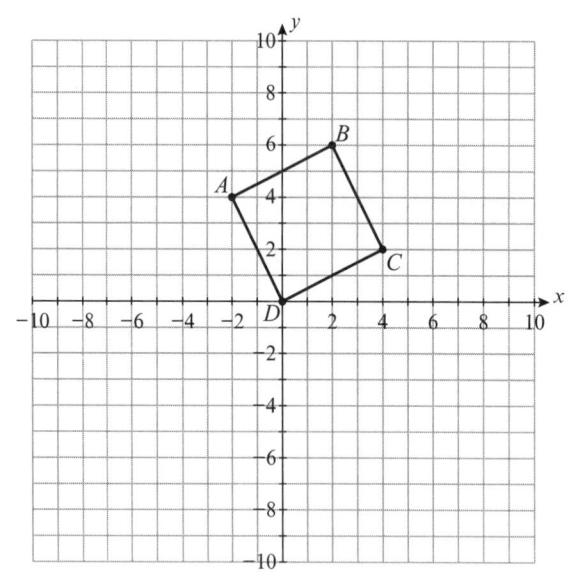

a Reflect *ABCD* in the line $x = -2$.

b Rotate *ABCD* 90° clockwise about the origin.

c Translate *ABCD* along the column vector $\begin{pmatrix} -3 \\ 2 \end{pmatrix}$.

d Enlarge *ABCD* by a scale factor of 1.5 using the origin as the centre of enlargement.

CONTINUED

4 $ABCD$ is shown on the grid. Make a copy of the diagram to work on.

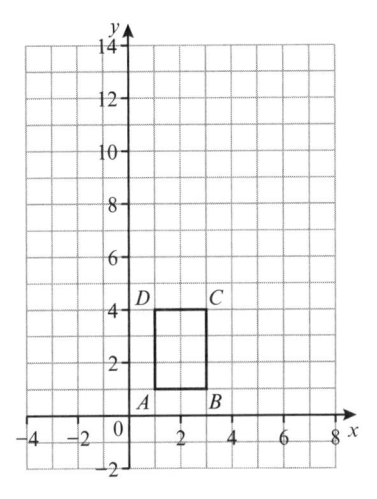

- **a** Enlarge $ABCD$ by a scale factor of 2 using $(0, 0)$ as the centre of enlargement. Label this $A'B'C'D'$.

- **b** On the same grid, enlarge $ABCD$ by a scale factor of 4 using $(2, 1)$ as the centre of enlargement. Label this $A''B''C''D''$.

- **c** Express the relationship between the length of AD and $A'D'$ as a ratio in simplest form.

- **d** Express the relationship between $A''B''$ and AB as a ratio in simplest form.

SELF ASSESSMENT

Think about how you would mark your own work in the Review Exercise.

Make a short list of the things you will check to make sure any transformations you draw are correct.

Use your list to assess your answers to questions 2, 3 and 4.

Is there anything you can improve? How will you do that?

Chapter 23: Probability using tree diagrams and Venn diagrams

23.1 Using tree diagrams to show outcomes

KEY LEARNING STATEMENTS

- A probability tree shows all the possible outcomes for two or more statistical experiments.
- Each line segment or branch represents one outcome. The end of each branch segment is labelled with the outcome and the probability of each outcome is written on the branch.

KEY CONCEPT

How to draw and label a tree diagram.

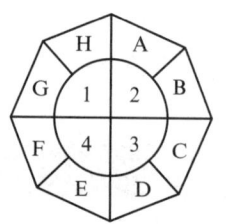

1. Anita has four cards. They are yellow, red, green and blue. She draws a card at random and then tosses a coin. Draw a tree diagram to show all possible outcomes.

2. The spinner shown has numbers on an inner circle and letters on an outer ring. When spun, it gives a result consisting of a number and a letter. Draw a tree diagram to show all possible outcomes when you spin it.

23.2 Calculating probability from tree diagrams

KEY LEARNING STATEMENTS

- To determine the probability of a combination of outcomes, multiply along each of the consecutive branches. If several combinations satisfy the same outcome conditions, then add the probabilities of the different paths.
- The sum of all the probabilities on a set of branches must equal one.

KEY CONCEPT

Using the probabilities on the branches to work out the probability of combined events.

23 Probability using tree diagrams and Venn diagrams

1. When a coloured ball is drawn from a bag, a coin is tossed once or twice, depending on the colour of the ball drawn. There are three blue balls, two yellow balls and a black ball in the bag. The tree diagram shows the possible outcomes.

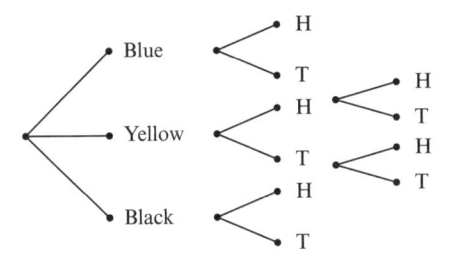

   a Copy and label the diagram to show the probability of each outcome. Assume the draw of the balls is random and the coin is fair.

   b Calculate the probability of a blue ball and a head.

   c Calculate the probability of a yellow ball and two heads.

   d Calculate the probability that you will not get heads at all.

2. The tree diagram below shows the possible outcomes when two coins are tossed.

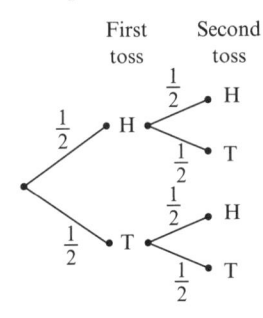

   a Copy and complete the tree diagram to show the possible outcomes when a third coin is tossed.

   b Calculate the probability of tossing three heads.

   c Calculate the probability of getting at least two tails.

   d Calculate the probability of getting fewer heads than tails.

   e Calculate the probability of getting an equal number of heads and tails.

3. When Zara drives to work she goes through a set of traffic lights and she passes a pedestrian crossing. She has worked out that the probability of the traffic lights being green is $\frac{3}{4}$ and the probability that she has to stop for a pedestrian is $\frac{2}{9}$.

   a Draw a tree diagram to represent this situation.

   b Calculate the probability that the traffic lights are not green when Zara gets to them and she has to stop at the pedestrian crossing.

   c What is the probability that the traffic lights are green and she does not have to stop for a pedestrian?

23.3 Calculating probability from Venn diagrams

KEY LEARNING STATEMENTS

- Venn diagrams are useful for working out probability problems, especially if the information involves the intersection or union of outcomes.

- The probability of two events happening is the same as the probability that an element is in both set A and set B. This is the intersection of set A and set B in a Venn diagram. The word 'and' is a clue that the probability is found in the intersection of the sets.

- The probability of either event A happening or event B happening is the same as the probability that an element is found either in set A or in set B. This is the union of set A and set B in the Venn diagram. The word 'or' is a clue that the probability is found in the union of the sets.

KEY CONCEPT

Calculating probability of combined events using Venn diagrams.

1. Josh has 12 identical cards with the numbers 1 to 12 written on them. He shuffles them and chooses a card at random. By drawing a Venn diagram, find the probability that the number on the card is:

   a even

   b an even number or a multiple of 3

   c an even multiple of 3

   d neither even, nor a multiple of 3.

2. A group of 75 students have a choice of two sports: basketball or swimming. 33 students play basketball and 47 students swim. 24 students do both sports.

   a Represent this data on a Venn diagram.

   b Work out the probability that a student chosen at random from this group will:

       i   do both sports   ii   do neither sport

       iii do at least one of the sports.

REFLECTION

If you practise solving lots of different probability problems you will be able to tackle this topic confidently.

- Different types of problems require different approaches. What helps you decide how to tackle a particular problem?

- What additional resources can you use to for additional practice and to challenge yourself? List these and describe how each one is useful.

REVIEW EXERCISE

1 Dineo is playing a game where she rolls a normal six-sided dice and tosses a coin. If the score on the dice is even, she tosses the coin once. If the score on the dice is odd, she tosses the coin twice.

    a Draw a tree diagram to show all possible outcomes.

    b Assuming the dice and the coin are fair and all outcomes are equally likely, label the branches with the correct probabilities.

    c What is the probability of obtaining two tails?

    d What is the probability of obtaining a five, a head and a tail (in any order)?

2 Zarah has 20 markers in her pencil case. There are six whiteboard markers and four green markers. Only one of the whiteboard markers is green.

    a Draw a Venn diagram to show this information.

    b Work out the probability that a marker chosen at random is:

        i not green

        ii a non-green whiteboard marker

        iii neither green nor a whiteboard marker.

3 Mrs Chan has a choice of ten mobile phone packages. Six of the packages offer free data bundles, five offer a free hands-free kit and three offer both.

    a Draw a Venn diagram to show this information.

    b Mrs Chan decides she is going to pick a package at random. Determine the probability of:

        i getting free data only

        ii getting free data and a hands-free kit

        iii getting free data or a hands-free kit.

    c Express her chances of getting neither as a percentage.

> Acknowledgements

The authors and publishers acknowledge the following sources of copyright material and are grateful for the permissions granted. While every effort has been made, it has not always been possible to identify the sources of all the material used, or to trace all copyright holders. If any omissions are brought to our notice, we will be happy to include the appropriate acknowledgements on reprinting.

Thanks to the following for permission to reproduce images:

Cover Wasan Prunglampoo/Getty Images

Inside Unit 7 Jeremy Hudson/Getty Images; *Unit 15* HRAUN/Getty Images